CORONEL **SERIES** 03

EDITORIAL

INFINITO

[∞]

Making history!

www.editorialinfinito.com.ve

250 SOLVED EXERCISES OF INDEFINITE INTEGRALS

(c) Editorial Infinito, 2019

(c) Pedro Pablo CORONEL PÉREZ / Pablo Josué CORONEL LÓPEZ

LEGAL DEPOSIT: lf07620153703627

ISBN: 978-980-18-0388-1

INTERNAL LAYOUT: Editorial Infinito / Estudiográfico

COVER DESIGN: Estudiográfico (2019)

LITERARY EDITOR: Magister / Lcdo. Pedro Alberto Coronel López

TRANSCRIPTION ADVISOR: Prof. Néstor Fidel Herrera Colmenares

DIGITAL PRINTING: Editorial Infinito, San Cristóbal

Any observation, suggestions and correspondence are kindly requested to be sent to the following emails: **pedro_coronel1955@hormail.com / pablocoronel19@hotmail.com**

Produced by Editorial Infinito (Bolivarian Republic of Venezuela)

250

SOLVED EXERCISES OF INDEFINITE
INTEGRALS

[INCLUDES THEORETICAL BASIS]

• Exercises solved step by step • Recommended for self-directed courses
• A very useful tool for both students and teachers
• Ideal for an efficient preparation for exams • Strengthens math skills
in problems solving

PEDRO P. **CORONEL PÉREZ** | PABLO J. **CORONEL LÓPEZ**

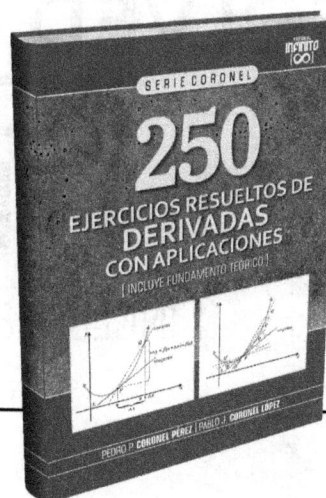

To **Srinivasa Ramanujan**, Indian self-taught mathematician

*"An equation has no meaning for me unless
I express a thought of God."*
Srinivasa Ramanujan (1887-1920) / Indian self-taught mathematician

CONTENTS

PEDRO PABLO CORONEL PÉREZ

Retired professor of the university institute of agroindustrial technology Region los Andes Bolivarian Republic of Venezuela. He was assigned to the department of electronics in electrical circuits and physics laboratory. I collaborate in the national training project in the area of mathematics..

PABLO JOSUÉ CORONEL LÓPEZ

Electrical engineer graduated from the university institute of agroindustrial technology Region los Andes, Bolivarian Republic of Venezuela.

COMMENTS FROM THE AUTHORS

Welcome to the first edition of **250 solved exercises of indefinite Integrals**! The Editorial Infinito proudly presents the aforementioned book to the student and teacher community. The purpose of this book is to present to those who start university studies, a series of exercises on indefinite integrals, very representative and resolved in detail. Obviously, it will be very useful for career students linked to engineering, science, technology or any specialty where mathematical calculation is an essential requirement within the study curriculum.

The number of exercises included allows the book can also be used as text by both the student and the teacher in the development of this important subject of calculation.

The authors have been careful in the explanation of the procedures used in the resolution of each of the problems. The exercises have been selected in order to expand the knowledge acquired in class by the alumni, as well as for the student to acquire practice in solving problems and thus prevent the difficulties with which the beginner normally stumbles.

It is expected that you will enjoy the first edition of 250 Resolved Exercises of indefinite Integrals. As always, comments and suggestions are welcome so we can continue improving our work in the next books to come.

Pedro Pablo Coronel Pérez
AUTHOR

Pablo Josué Coronel López
AUTHOR

Pedro Alberto Coronel López
LITERARY ADVISOR

THEORETICAL BASIS
FOR INDEFINITE INTEGRATION

ANTI DERIVED OR PRIMITIVE

In the theoretical basis for the derivation reference was made only to the following basic problem: Given a function **f** find its derivative **f'**. In the present theoretical foundation of this book, it will be seen that an equally important problem is:

Given a function **f**, find a function whose derivative is the given **f**.

That is, for a given function **f**, you want to find another function F for which $F'(x) = f(x)$ for all x in a certain interval.

Definition

It is said that a function F is an antiderivative or primitive of a function **f** if $F'(x) = f(x)$ in some interval.

Example:

An antiderivative or primitive of $f(x) = 2x$ is $F(x) = x^2$, since $F'(x) = 2x$.

There is always more than one antiderivative of a function. In the case of the previous example, $F_1(x) = x^2 - 1$ y $F_2(x) = x^2 - 10$ are also antiderivatives f $f(x) = 2x$, since $F_1'(x) = F_2'(x) = f(x)$ in effect, if F is an antiderivative of a function **f** then $G(x) = F(x) + C$ is also, for any constant C. This is a consequence of the fact that:

$$G'(x) = \frac{d}{dx}(F(x) + C) = F'(x) + 0 = F'(x) = f(x).$$

Then, $F(x) + C$ represents the family of antiderivatives of which each member has a derivative equal to $f(x)$. Where $F(x) + C$ is the most general antiderivative of $f(x)$.

NOTATION OF THE INDEFINITE INTEGRAL

For convenience we introduce a notation for an antiderivative or primitive of a function. If $F'(x) = f(x)$, the most general antiderivative of **f** will be represented by:

$$\int f(x)dx = F(x) + C$$

The symbol $\int$ is called the integral symbol, and the notation $\int f(x)dx$ is called the indefinite integral of $f(x)$ with respect to x. The function $f(x)$ is called integrand. The process of finding an antiderivative or primitive is called antidifferentiation or intricacy. The number C is known as the integration constant. Just as the symbol d/dx ( ) denotes differentiation with respect to x, the symbol $\int f(x)dx$ denotes integration with respect to x.

THE INDEFINITE INTEGRAL OF A POWER

By differentiating the power x^n, the exponent n is set as a factor and the value of the original exponent is decreased by 1. To find an antiderivative or primitive of x^n , the opposite of the differentiation rule would be: increase the exponent by 1 and divide by the new exponent $n + 1$. The analogous rule, for the indefinite integral, according to the rule of differentiation of a power, is as follows:

If n is a rational number, then for $n \neq -1$

$$\int x^n\ dx = \frac{x^{n+1}}{n+1} + c$$

Example 2

Evaluate $\int x^6\,dx$

With n = 6, and applying the indefinite integral of a power you have

$$\int x^6\,dx = \frac{x^{6+1}}{6+1} + c = \frac{x^7}{7} + c$$

The next property of indefinite integrals is an immediate consequence of the fact that the derivative of a sum is the sum of the derivatives.

Theorem

$$F'(x) = f(x) \ \ y \ \ G'(x) = g(x)\, , so$$

$$\int [f(x) \pm g(x)]dx = \int f(x)dx \pm \int g(x)dx = F(x) \pm G(x) + C$$

Note that there is no reason to use two integration constants, since

$$\int [f(x) \pm g(x)]dx = (F(x) + C_1) \pm (G(x) + C_2) = F(x) \pm G(x) + (C_1 \pm C_2)$$

$$= F(x) \pm G(x) + C$$

en donde se ha reemplazado $C_1 \pm C_2$ por la constante única C.

RESOLUTION OF A DIFFERENTIAL EQUATION

A differential equation in x and y is an equation that includes x and y to the derivatives of Y. FOR EXAMPLE, $y'=3x$ y $y'=x^2+1$ they are examples of differential equations.

Find the general solution of the differential equation $y'=3$.

Solution: Initially, it consists in determining a function whose derivative is 3. A function of this characteristic is: $y=3x$,   *3x es una antiderivada de* **3**

Now, the general solution is:

$$y = 3x \longrightarrow General\ solution$$

When a differential equation of the form is solved

$$\frac{dy}{dx} = f(x)$$

It is convenient to write it in the equivalent differential form

$$dy = f(x)dx$$

The operation to find all the solutions of this equation is called antiderivation or indefinite integration. The general solution is denoted by:

$$y = \int f(x)dx = F(x) + C.$$

INITIAL CONDITIONS AND PARTICULAR SOLUTIONS

Let the following differential equation

$$\frac{dy}{dx} = 3x^2 - 1$$

When rewriting you have:

$$dy = (3x^2 - 1)dx$$

Where the general solution is:

$$y = \int (3x^2 - 1)dx$$

$$F(x) = x^3 - x + C \quad \rightarrow General\ solution$$

To determine a particular solution, consider that the curve passes through point **(2,4)**. This information is called the initial condition. Using this condition in the general solution, it is possible to determine: **F(2) = 8-2 + C = 4**, which means that **C = -2**. Therefore, we have:

$$F(x) = x^3 - x - 2 \quad \rightarrow Particular\ solution$$

The following table presents a summary of basic integration rules. Both for derivation and integration formulas.

BASIC RULES OF INTEGRATION

DERIVATION FORMULAS	INTEGRATION FORMULAS
$\frac{d}{dx}[C] = 0$	$\int 0\,dx = C$
$\frac{d}{dx}[kx] = k$	$\int k\,dx = kx + C$
$\frac{d}{dx}[kf(x)] = kf'(x)$	$\int kf(x)\,dx = k\int f(x)\,dx$
$\frac{d}{dx}[f(x) \pm g(x)] = f'(x) \pm g'(x)$	$\int [f(x) \pm g(x)]\,dx = \int f(x)\,dx \pm \int g(x)\,dx$
$\frac{d}{dx}[x^n] = nx^{n-1}$	$\int x^n\,dx = \frac{x^{n+1}}{n+1} + C,\ \ n \neq -1$
$\frac{d}{dx}[\operatorname{sen}x] = \cos x$	$\int \cos x\,dx = \operatorname{sen}x + C$

$\dfrac{d}{dx}[\cos x] = -\operatorname{sen} x$	$\displaystyle\int \operatorname{sen} x \, dx = -\cos x + C$
$\dfrac{d}{dx}[\tan x] = \sec^2 x$	$\displaystyle\int \sec^2 x \, dx = \tan x + C$
$\dfrac{d}{dx}[\sec x] = \sec x \tan x$	$\displaystyle\int \sec x \tan x \, dx = \sec x + C$
$\dfrac{d}{dx}[\cot x] = -\csc^2 x$	$\displaystyle\int \csc^2 x \, dx = -\cot x + C$
$\dfrac{d}{dx}[\csc x] = -\csc x \cot x$	$\displaystyle\int \csc x \cot x \, dx = -\csc x + C$

INTEGRATION BY SUBSTITUTION (CHANGE OF VARIABLE)

With a change of formal variables, the integral can be completely rewritten in terms of **u** and **du** (or any other convenient variable). The variable change technique uses Leibniz's notation for the differential. That is, if **u** = **g** (**x**), then **du** = **g′**(**x**)**dx**.

Theorem

Let **g** be a function whose path or range is a range **I**, and let **f** be a continuous function in **I**. If **g** is derivable in its domain and **F** is an antiderivative or primitive of **f** in **I**, then:

$$\int f(g(x))g'(x)dx = F(g(x)) + C$$

Si $u = g(x)$, so $du = g'(x)dx$ y

$$\int f(u) \, du = F(u) + C$$

Strategies to make a change of variable

Choose a substitution u = g (x). Usually, it is better to choose the internal part of a composite function, such as a quantity raised to a power.

1. Calculate du = g′(x) dx.
2. Rewrite the integral in terms of the variable u.
3. Find the resulting integral in terms of u.
4. Replace u with g (x) to obtain an antiderivative or primitive in terms of x.

5. Verify the answer by derivation.

Integration by parts

In this area of study, an important integration technique called integration by parts will be analyzed. This technique can be applied to a wide variety of functions and is particularly useful for integrands that contain products with algebraic and transcendental functions. For example, piecemeal integration works well with integrals like

$$\int x \, lnx \, dx, \int x^2 e^x dx \;\; y \;\; \int e^x sen \, xdx$$

To deduce the formula that corresponds to the integration by parts, the following should be taken into account: Every derivation rule has a corresponding integration rule. The rule that corresponds to the rule of the product for derivation is called rule for integration by parts.

It is known that the derivative of a product between two functions is

$$d(uv) = udv + vdu$$

Where **u** and **v** are derivable functions of x.

By integrating both sides of the equation you have:

$$\int d(uv) = \int udv + \int vdu$$

$$uv = \int udv + \int vdu$$

Clearing $\int udv$

$$\boxed{\int udv = uv - \int vdu}$$

This formula expresses the original integral (integral to be resolved) in terms of another integral.

One of the situations that readers ask themselves is: How do I learn the formula? Well there are a series of expressions that allow us to remember the formula, one of them our favorite is: **One Day I saw a brave** soldier **dressed in uniform**. The idea is to keep the first letter (uppercase) to reconstruct the formula.

Another situation that is often presented to readers is the selection of both u and dv. Since depending on the selection of u and dv it may be easier to evaluate the second integral than the original. Mathematicians have defined rules that allow us to have a guide for the selection of u and dv. One of them is the ALPES rule.

What does this rule consist of? First, let's discover what is meant by each of the letters of this word

- A: Arc functions (sine arc, cosine arc, tangent arc)
- L: Logarithms
- P: Powers (of numerical exponent)
- E: Exponential
- S: Sine and cosine

Well, how do we use all this? Very easy:

It is convenient to use the integration method by parts when we have an integral of an arc function only, a logarithm only or a product of two functions that belong to two of those five types.

In the first case, only one arc function, we will call that arc function and the rest (in this case); in the second case, only one logarithm, we will call the logarithm and the rest (also); and in the third case, the most interesting, that of the product, we will call the function whose type appears first in ALPES and the rest (which will now be the other function by). For example, the integral

$$\int x \log(x)dx$$

is a product of, that belongs to P, and $\log(x)$, who enters L. As in ALPES the L appears before the P, the assignment will be

$$u = \log(x) \qquad dv = x\, dx$$

There is another rule like ILATE that also works. In this book, the ALPES rule will be used. There are cases in which it is useless, since the function to integrate does not have elementary primitive, and in other cases it is necessary to be careful, very careful, when applying the method.

TRIGONOMETRIC INTEGRALS

Integrals that contain sine and cosine powers

In this section we will study the techniques to evaluate integrals of the types

$$\int sen^m x \, cos^n x \, dx \quad y \quad \int sec^m x \, tan^n x \, dx$$

Where **m** or **n** is any positive integer. To find the antiderivative or primitive for these expressions, try to break them into combinations of integrals to which the rule of powers can be applied. To separate $\int sen^m x \, cos^n x \, dx$ in ways to which the power rule can be applied, use the following identities

$$sen^2 x + cos^2 x = 1 \quad Pythagorean\ identity$$

$$sen^2 x = \frac{1 - \cos 2x}{2} \quad Identity\ of\ the\ average\ angle\ for\ sen^2 x$$

$$cos^2 x = \frac{1 + \cos 2x}{2} \quad Identity\ of\ the\ average\ angle\ for\ cos^2 x$$

Next, a set of strategies that allow evaluating integrals containing sines and cosines are presented.

1. If the sine power is odd and positive, keep a sine factor and pass the remaining factors to cosines. Then, it is operated and integrated.

2. If the cosine power is odd and positive, keep a cosine factor and pass the remaining factors to sines. Then, it is operated and integrated.

3. If the powers of both are even and not negative, use identities repeatedly

$$sen^2 x = \frac{1 - \cos 2x}{2} \quad y \quad cos^2 x = \frac{1 + \cos 2x}{2}$$

To convert the integrand to odd powers of the cosine. Then proceed as in strategy 2.

Integrals that contain drying and tangent powers

Next, a set of strategies that allow to evaluate integrals of the form are presented:

$$\int \sec^m x \, \tan^n x \, dx$$

1. If the power of the secant is even and positive, keep a square secant factor and pass the remaining factors to tangents. Then, it is operated and integrated.

2. If the power of the secant is odd and positive, keep a drying tangent factor and convert the remaining factors to drying. Then, it is operated and integrated.

3. If there are no drying factors and the tangent power is even and positive, convert a square tangent factor to square secant.

4. If the integral is of the form $\int \sec^m x \, dx$ where **m** is odd and positive, use integration by parts.

5. If none of the first four guides applies, try to convert the integrand into sines and cosines.

INTEGRATION BY TRIGONOMETRIC SUBSTITUTION

The integrals that contain radicals of the form

$$\sqrt{a^2 - u^2}, \sqrt{a^2 + u^2}, \sqrt{u^2 - b^2}$$

it is possible that they can be evaluated by means of a trigonometric substitution. The objective of the trigonometric substitution is to eliminate the integrand radical. To achieve the above, Pythagorean identities are required.

$$cos^2\theta = 1 - sen^2\theta, \quad sec^2\theta = 1 + tan^2\theta \quad y \quad tan^2\theta = sec^2\theta - 1$$

Trigonometric substitutions (a>0)

Case I Integrals that contain $\sqrt{a^2 - u^2}$

It does $= a \, sen \, \theta \, , \, -\dfrac{\pi}{2} \le \theta \le \dfrac{\pi}{2}$

It is replaced, $u = a\,sen\,\theta$ in the radical

$$\sqrt{a^2 - u^2} = \sqrt{a^2 - a^2 sen^2\theta}$$

$$= \sqrt{a^2(1 - sen^2\theta)}$$

$$= \sqrt{a^2 cos^2\theta}$$

$$= a\cos\theta$$

Therefore, we have: $\sqrt{a^2 - u^2} = a\cos\theta$

From the above it can be inferred, that if an integral intervenes an algebraic term $\sqrt{a^2 - u^2}$ it becomes a trigonometric integral. After integration, the variable θ can be eliminated by using a right triangle. Since $u = a\,sen\theta$ and clearing $sen\theta = u/a$ the right triangle is obtained. Hence the importance of the construction of the triangle since it allows us to return to the original variable.

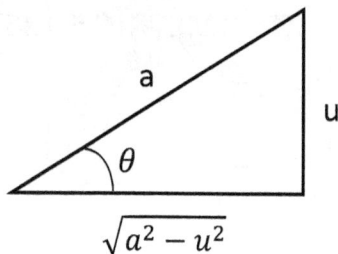

Case II Integrals that contain $\sqrt{a^2 + u^2}$

It does $= a\,tan\,\theta\,,\, -\dfrac{\pi}{2} < \theta < \dfrac{\pi}{2}$

It is replaced $u = a\,tan\,\theta$ in the radical:

$$\sqrt{a^2 + u^2} = \sqrt{a^2 + a^2 tan^2\theta}$$

$$= \sqrt{a^2(1 + tan^2\theta)}$$

$$= \sqrt{a^2 sec^2\theta} = a \sec \theta$$

Therefore, we have: $\sqrt{a^2 + u^2} = a \sec \theta$

From the above it can be inferred, that if an integral intervenes an algebraic term $\sqrt{a^2 + u^2}$ it becomes a trigonometric integral. After integration, the variable θ can be eliminated by using a right triangle. As $u = a \ tan \ \theta$ and clearing $tan \ \theta = u/a$ you get the right triangle. Hence the importance of the construction of the triangle since it allows us to return to the original variable.

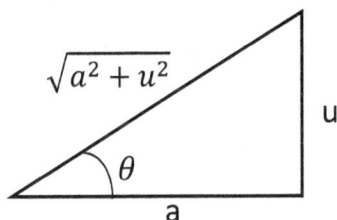

Case III ntegrals that contain $\sqrt{u^2 - a^2}$

It does $= a \ sec \ \theta$, $0 \le \theta < \frac{\pi}{2}$, ó $\pi \le \theta < \frac{3\pi}{2}$

It is replaced $u = a \ sec \ \theta$ in the radical

$$\sqrt{u^2 - a^2} = \sqrt{a^2 sec^2\theta - a^2}$$

$$= \sqrt{a^2(sec^2\theta - 1)}$$

$$= \sqrt{a^2 tan^2\theta}$$

$$= a \tan \theta$$

Therefore, we have: $\sqrt{u^2 - a^2} = a \tan \theta$

From the above it can be inferred, that if an integral intervenes an algebraic term $\sqrt{u^2 - a^2}$ it becomes a trigonometric integral. After integration, the variable can be eliminated θ using a right triangle. As $u = a \ sec \ \theta$ and clearing $sec \ \theta = u/a$ you get the right triangle. Hence the importance of the construction of the triangle since it allows us to return to the original variable.

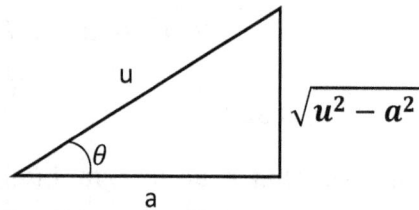

INTEGRATION OF RATIONAL FUNCTIONS

To integrate certain rational functions such as $\frac{P(x)}{Q(x)}$, where the degree of $P(x)$ is less than the degree of $Q(x)$ the known method is that of partial fractions. That consists of decomposing said rational function into simpler component fractions, and then evaluating the integral term by term. In this section, four cases of partial fraction decomposition will be studied.

Denominators that contain linear factors

Case I Non-repeated linear factors
The following algebraic result is established, without demonstration. If

$$\frac{P(x)}{Q(x)} = \frac{P(x)}{(a_1x + b_1)(a_2x + b_2). \ . \ . \ (a_nx + b_n)}$$

where all the factors $a_i x + b_i$, $i = 1, 2,...,$ **n** are different and the degree of $P(x)$ is less than **n**, so there are only real constants $C_1, C_2,..., C_n$ such that

$$\frac{P(x)}{Q(x)} = \frac{C_1}{a_1x + b_1} + \frac{C_2}{a_2x + b_2} + . \ . \ . + \frac{C_n}{a_nx + b_n}$$

Case II Repeated linear factors

If,
$$\frac{P(x)}{Q(x)} = \frac{P(x)}{(ax + b)^n}$$

Where **n > 1** and the degree of $P(x)$ is less than **n**, then you can find unique real constants $C_1, C_2,..., C_n$ such that

$$\frac{P(x)}{(ax+b)^n} = \frac{C_1}{ax+b} + \frac{C_2}{(ax+b)^2} + \ldots + \frac{C_n}{(ax+b)^n}$$

Denominators that contain irreducible quadratic factors

Case III Non-repeated quadratic factors
Suppose that the denominator of the rational function $\frac{P(x)}{Q(x)}$, can be expressed as a product of different irreducible quadratic factors $x^2 + b_i\, x + c$
$i = 1, 2,\ldots, n$. If, the degree of $P(x)$ is less than $2n$, it is possible to find constants real
and uniques $A_1, A_2,\ldots, A_n, B_1, B_2,\ldots, B_n$ such that

$$\frac{P(x)}{Q(x)} = \frac{P(x)}{(a_1x^2 + b_1x + c_1)(a_2x^2 + b_2x + c_2)\ldots(a_nx^2 + b_nX + C_n)}$$

$$\frac{P(x)}{Q(x)} = \frac{A_1x + B_1}{a_1x^2 + b_1x + c_1} + \frac{A_2x + B_2}{a_2x^2 + b_2x + c_2} + \ldots + \frac{A_n + B_n}{a_nx^2 + b_nX + C_n}$$

Case IV Quadratic factors repeated
We now consider the case in which the integrand is $\frac{P(x)}{(ax^2 + bx + c)^n}$, where $a\,x^2 + bx + c$ it
is irreducible and $n > 1$. If the degree of $P(x)$ is less than $2n$, you can find unique real
constants $A_1, A_2,\ldots, A_n, B_1, B_2,\ldots, B_n$ such that

$$\frac{P(x)}{(ax^2 + bx + c)^n} = \frac{A_1x + B_1}{ax^2 + bx + c} + \frac{A_2x + B_2}{(ax^2 + bx + c)^2} + \ldots + \frac{A_n + B_n}{(ax^2 + bx + c)^n}$$

INTEGRATION OF RATIONAL SINE AND COSINE FUNCTIONS

Integrals of rational expressions in which $sin\ x$ and $cos\ x$ can be reduced to integral
polynomial ratios by means of substitution

$u = tan\frac{x}{2}$, si $\frac{x}{2}$ represents the angle shown in the figure

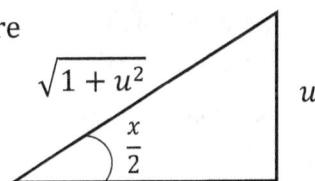

25

$$\text{Then,} \qquad sen\frac{x}{2} = \frac{u}{\sqrt{1+u^2}} \qquad y \qquad \cos\frac{x}{2} = \frac{1}{\sqrt{1+u^2}}$$

The trigonometric identities for double angles have,

$$sen\, 2x = 2\, sen\, x\, \cos x$$

$$sen\, 2\frac{x}{2} = 2\, sen\frac{x}{2}\cos\frac{x}{2} = 2\frac{u}{\sqrt{1+u^2}}\frac{1}{\sqrt{1+u^2}} \Rightarrow \boxed{sen\, x = \frac{2u}{1+u^2}}$$

$$\cos 2x = \cos^2 x - sen^2 x$$

$$\cos 2\frac{x}{2} = \cos^2\frac{x}{2} - sen^2\frac{x}{2} = \left(\frac{1}{\sqrt{1+u^2}}\right)^2 - \left(\frac{u}{\sqrt{1+u^2}}\right)^2 = \frac{1}{1+u^2} - \frac{u}{1+u^2}$$

$$\boxed{\cos x = \frac{1-u^2}{1+u^2}}$$

As

$$u = tan\frac{x}{2} \rightarrow du = sec^2\frac{x}{2}\frac{dx}{2}$$

$$du = \left(1 + tan^2\frac{x}{2}\right)\frac{dx}{2} = (1+u^2)\frac{dx}{2} \Rightarrow \boxed{dx = \frac{2}{1+u^2}\, du}$$

BIBLIOGRAPHY

Larson R, Hostetler R. Y Edwards B. (1995). Calculation. Volume 1. México. McGraw-Hill.

Zill G, Dennis (1985). Calculation with Analytical Geometry. Mexico, Iberoamerica Editorial Group.

Wisniewski Piotr M. y Banegas G. Ana L. (2004). Introduction to university mathematics. Mexico. McGraw-Hill.

Pita R, Claudio. (1998). Calculation of a variable. Mexico, Prentice- Hispano-American Hall.

**In the following exercises,
the integrals are found
and the results are verified by
means of the derivation.**

$$\int (x+3)\,dx$$

According to the theoretical foundation of this book, we have

$$\int (x+3)\,dx = \int x\,dx + \int 3\,dx$$

For the first integral the power rule is applied and for the second the rule of the constant multiple (see summary table of basic rules of integration in the theoretical basis of this book).

$$\int (x+3)\,dx = \int x\,dx + 3\int dx = \frac{x^{1+1}}{1+1} + 3x + C$$

$$\int (x+3)\,dx = \frac{x^2}{2} + 3x + C$$

$$\boxed{\int (x+3)\,dx = \frac{1}{2}x^2 + 3x + C}$$

Now we proceed to verify the result through the derivation

$$\frac{d}{dx}\left[\frac{1}{2}x^2 + 3x + C\right] = \frac{1}{2}\frac{d}{dx}[x^2] + 3\frac{d}{dx}[x] + \frac{d}{dx}[c] = x + 3$$

The result is verified, since

$$\frac{d}{dx}\left[\frac{1}{2}x^2 + 3x + C\right] = x + 3 = f(x)$$

002

$$\int \frac{x^2 + x + 1}{\sqrt{x}}\, dx$$

The integral is rewritten

$$\int \frac{x^2 + x + 1}{x^{\frac{1}{2}}}\, dx = \int \left(\frac{x^2}{x^{\frac{1}{2}}} + \frac{x}{x^{\frac{1}{2}}} + \frac{1}{x^{\frac{1}{2}}} \right) dx = \int \left(x^{\frac{3}{2}} + x^{\frac{1}{2}} + x^{-\frac{1}{2}} \right) dx$$

$$\int \frac{x^2 + x + 1}{x^{\frac{1}{2}}}\, dx = \int x^{\frac{3}{2}}dx + \int x^{\frac{1}{2}}dx + \int x^{-\frac{1}{2}}dx = \frac{x^{\frac{5}{2}}}{\frac{5}{2}} + \frac{x^{\frac{3}{2}}}{\frac{3}{2}} + \frac{x^{\frac{1}{2}}}{\frac{1}{2}} + c$$

$$\int \frac{x^2 + x + 1}{x^{\frac{1}{2}}}\, dx = \frac{2}{5}x^{\frac{5}{2}} + \frac{2}{3}x^{\frac{3}{2}} + 2x^{\frac{1}{2}} + c = \frac{6x^{\frac{5}{2}} + 10x^{\frac{3}{2}} + 30x^{\frac{1}{2}}}{15} + c$$

$$\boxed{\int \frac{x^2 + x + 1}{\sqrt{x}}\, dx = \frac{2}{15}x^{\frac{1}{2}}(3x^2 + 5x + 15) + c}$$

Now we proceed to verify the result through the derivation. But before the indicated product is made

$$\frac{2}{15}x^{\frac{1}{2}}(3x^2 + 5x + 15) + c = \frac{2}{5}x^{\frac{5}{2}} + \frac{2}{3}x^{\frac{3}{2}} + 2x^{\frac{1}{2}} + c$$

$$\frac{d}{dx}\left[\frac{2}{5}x^{\frac{5}{2}} + \frac{2}{3}x^{\frac{3}{2}} + 2x^{\frac{1}{2}} + c \right] = \frac{10}{10}x^{\frac{3}{2}} + \frac{6}{6}x^{\frac{1}{2}} + \frac{2}{2}x^{-\frac{1}{2}} = x^{\frac{3}{2}} + x^{\frac{1}{2}} + x^{-\frac{1}{2}}$$

$$\frac{d}{dx}\left[\frac{2}{5}x^{\frac{5}{2}} + \frac{2}{3}x^{\frac{3}{2}} + 2x^{\frac{1}{2}} + c \right] = \frac{x^{\frac{1}{2}}\left(x^{\frac{3}{2}} + x^{\frac{1}{2}} + x^{-\frac{1}{2}} \right)}{x^{\frac{1}{2}}}$$

It is verified that

$$\frac{d}{dx}\left[\frac{2}{5}x^{\frac{5}{2}} + \frac{2}{3}x^{\frac{3}{2}} + 2x^{\frac{1}{2}} + c \right] = \frac{x^2 + x + 1}{\sqrt{x}} = f(x)$$

$$\int (x+1)(3x-2)dx$$

003

The integral is rewritten

$$\int (x+1)(3x-2)dx = \int (3x^2+x-2)dx$$

$$\int (3x^2+x-2)dx = 3\int x^2 dx + \int x dx - 2\int dx$$

$$\int (3x^2+x-2)dx = 3\frac{x^3}{3} + \frac{x^2}{2} - 2x + c = x^3 + \frac{1}{2}x^2 - 2x + c$$

$$\boxed{\int (x+1)(3x-2)dx = x^3 + \frac{1}{2}x^2 - 2x + c}$$

Now we proceed to verify the result through the derivation

$$\frac{d}{dx}\left[x^3 + \frac{1}{2}x^2 - 2x + c\right] = 3x^2 + x - 2$$

Then the trinomial is factored, multiplying it by the coefficient of x^2

$$3x^2 + x - 2 = 9x^2 + 3x - 6$$

Now it is written as follows

$$9x^2 + 3x - 6 = (3x)^2 + (3x) - 6 = (3x+3)(3x-2)$$

As the trinomial multiplied by 3, it is now divided by 3

$$\frac{(3x+3)(3x-2)}{3} = (x+1)(3x-2)$$

It is verified that

$$\frac{d}{dx}\left[x^3 + \frac{1}{2}x^2 - 2x + c\right] = (x+1)(3x-2) = f(x)$$

004

$$\int (2t^2 - 1)^2 dt$$

The integral is rewritten

$$\int (2t^2 - 1)^2 dt = \int (4t^4 - 4t^2 + 1)dt$$

$$\int (4t^4 - 4t^2 + 1)dt = 4\int t^4 dt - 4\int t^2 dt + \int dt$$

$$\int (4t^4 - 2t^2 + 1)dt = 4\frac{t^5}{5} - 4\frac{t^3}{3} + t + C = \frac{4}{5}t^5 - \frac{4}{3}t^3 + t + c$$

$$\boxed{\int (2t^2 - 1)^2 dt = \frac{4}{5}t^5 - \frac{4}{3}t^3 + t + c}$$

Now we proceed to verify the result through the derivation

$$\frac{d}{dt}\left[\frac{4}{5}t^5 - \frac{4}{3}t^3 + t + c\right] = \frac{20}{5}t^4 - \frac{12}{3}t^2 + 1 = 4t^4 - 4t^2 + 1$$

To factor the previous expression, it is necessary to rewrite it

$$4t^4 - 4t^2 + 1 = (2t^2)^2 - 4t^2 + 1$$

Resulting in a perfect square trinomial, and its factorization is

$$(2t^2)^2 - 4t^2 + 1 = (2t^2 - 1)^2$$

It is verified that:

$$\frac{d}{dt}\left[\frac{4}{5}t^5 - \frac{4}{3}t^3 + t + c\right] = (2t^2 - 1)^2 = f(t)$$

$$\int (4x^3 + 6x^2 - 1)\, dx$$

005

$$\int (4x^3 + 6x^2 - 1)dx = 4\int x^3 dx + 6\int x^2 dx - \int dx$$

$$\int (4x^3 + 6x^2 - 1)dx = 4\frac{x^4}{4} + 6\frac{x^3}{3} - x + c = x^4 + 2x^3 - x + c$$

$$\boxed{\int (4x^3 + 6x^2 - x)\, dx = x^4 + 2x^3 - x + c}$$

Now we proceed to verify the result through the derivation

$$\frac{d}{dx}[x^4 + 2x^3 - x + c] = 4x^3 + 6x^2 - 1$$

It is verified that:

$$\frac{d}{dx}[x^4 + 2x^3 - x + c] = 4x^3 + 6x^2 - 1 = f(x)$$

For the resolution of integral exercises like the previous ones, keep in mind the next information:

It is said that a function F is an antiderivative or primitive of a function f if $F'(x) = f(x)$ in some interval.

006

$$\int (2sen\,x + 3\cos x)dx$$

$$\int (2sen\,x + 3\cos x)dx = 2\int senx\,dx + 3\int cosx\,dx$$

$$\int (2sen\,x + 3\cos x)dx = -\,2\cos x + 3sen\,x + c$$

$$\boxed{\int (2sen\,x + 3\cos x)dx = -2\cos x + 3sen\,x + c}$$

Now we proceed to verify the result through the derivation

$$\frac{d}{dx}[-2\cos x + 3sen\,x + c] = 2sen\,x + 3\cos x$$

It is verified that

$$\frac{d}{dx}[-2\cos x + 3sen\,x + c] = 2sen\,x + 3\cos x = f(x)$$

007

$$\int \sec y(\tan y - \sec y)\,dy$$

The integral is rewritten

$$\int \sec y(\tan y - \sec y)\,dy = \int (\sec y\tan y - \sec^2 y)dy$$

$$\int (\sec y\tan y - \sec^2 y)dy = \int \sec y\tan y\,dy - \int \sec^2 y\,dy$$

$$\boxed{\int \sec y(\tan y - \sec y)\,dy = \sec y - \tan y + c}$$

Now we proceed to verify the result through the derivation

It is verified that $\frac{d}{dy}[\sec y - \tan y + c] = \sec y\tan y - \sec^2 y = f(y)$

$$\int \frac{\cos x}{1 - \cos^2 x}\, dx$$

008

The integral is rewritten taking into consideration the Pythagorean trigonometric identity
$sen^2 x + \cos^2 x = 1$

$$\int \frac{\cos x}{1 - \cos^2 x}\, dx = \int \frac{\cos x}{sen^2 x}\, dx = \int \frac{\cos x}{senx}\frac{1}{sen\, x}\, dx = \int \cot x \csc x\, dx$$

Then proceed to integrate (see basic rules of integration)

$$\int \cot x \csc x\, dx = -\csc x + c$$

$$\boxed{\int \frac{\cos x}{1 - \cos^2 x}\, dx = -\csc x + C}$$

Now we proceed to verify the result through the derivation

It is verified that

$$\frac{d}{dx}[-\csc x + c] = -(-\csc x\, \cot x) = \csc x \cot x$$

$$\csc x \cot x = \frac{1}{sen\, x}\frac{\cos x}{sen\, x} = \frac{\cos x}{sen^2 x} = \frac{\cos x}{1 - \cos^2 x} = f(x)$$

009

$$\int (t^2 - sen\, t)\, dt$$

$$\int (t^2 - sen\, t)\, dt = \int t^2\, dt - \int sen\, t\, dt = \frac{t^3}{3} + \cos t + c$$

$$\boxed{\int (t^2 - \boldsymbol{sen\, t})\, \boldsymbol{dt} = \frac{1}{3}t^3 + \cos t + c}$$

Now we proceed to verify the result through the derivation
It is verified that

$$\frac{d}{dt}\left[\frac{1}{3}t^3 + \cos t + c\right] = t^2 - sen\, t = f(t)$$

010

$$\int (tan^2 y + 1)\, dy$$

$$\int (tan^2 y + 1)\, dy = \int tan^2 y\, dy + \int dy$$

Since the first integral is not immediate, it is rewritten taking into consideration the following trigonometric identity: $1 + tan^2 y = sec^2 y$

$$\int tan^2 y\, dy + \int dy = \int (sec^2 y - 1)dy + \int dy$$

$$\boxed{\int (\boldsymbol{tan^2 y + 1})\, \boldsymbol{dy} = \boldsymbol{tan\, y} + \boldsymbol{c}}$$

Now we proceed to verify the result through the derivation.

$$\frac{d}{dy}[tan\, y + \quad c] = sec^2 y = (tan^2 y + 1) = f(y)$$

**The following exercises show
how to use the integral calculus to analyze
problems such as: population growth,
shrub growth, vertical movement, lunar
gravity, among others.**

A nursery of green plants usually sell a certain shrub after 6 years of growth and care. The growth rate during those 6 years is, approximately, **dh/dt = 1.5t + 5**, where **t** is the time in years and **h** is the height in centimeters.
The seedlings are 12 centimeters tall when planted (t = 0).

a) Determine the height after t years
b) How tall are the bushes when they are sold?

Consider that t = 0 represents the initial time. The indicated initial condition can be written as follows: h (0) = 12

a) The speed of growth is rewritten as follows

$$\frac{dh}{dt} = 1.5t + 5 \quad \to \mathrm{dh} = (1.5t + 5)\mathrm{dt}$$

It integrates both members of the differential equation

$$\int dh = \int (1.5t + 5)dt = 1.5 \int t\, dt + 5 \int dt$$

$$h(t) = \frac{1.5}{2}t^2 + 5t + c \quad \to \quad h(t) = 0.75\, t^2 + 5t + c$$

$$h(0) = 0 + 0 + c = 12 \quad \to \boxed{h(t) = 0.75t^2 + 5t + 12}$$

b) With the above equation we have

$$h(6) = 0.75(6)^2 + 5*6 + 12 = 27 + 30 + 12 = 69cm.$$

$$\boxed{Shrubs\ are\ sold\ at\ a\ height\ of 69cm.}$$

012

The growth rate **dp/dt** of a population of bacteria is propor-
tional to the square root of **t**, where **p** is the size of the
population and **t** is the time in days (**0 ≤ t ≤ 10**). That is,
dp/dt= k√t. The initial population size is 500. After one day
the population has grown to 600. **Estimate the size of the
population after 7 days.**

Initial conditions presented by the exercise

For t = 0, the initial condition is written as follows: P (0) = 500
For t = 1 day, the information is: p (1) = 600

The growth rate is rewritten as follows

$$\frac{dp}{dt} = k\sqrt{t} \quad \rightarrow \quad dp = k\sqrt{t}\,dt$$

It integrates both members of the differential equation

$$\int dp = k\int \sqrt{t}\,dt = k\int t^{\frac{1}{2}}\,dt \rightarrow p(t) = k\,\frac{t^{\frac{3}{2}}}{\frac{3}{2}} + c = \frac{2}{3}k\,t^{\frac{3}{2}} + c$$

$$p(t) = \frac{2}{3}k\,t^{\frac{3}{2}} + c$$

To determine the value of C, it is taken into account that P (0) = 500

$$p(t) = \frac{2}{3}k\,t^{\frac{3}{2}} + c = p(0) = 0 + c = 500 \rightarrow c = 500$$

To determine the value of k, it is taken into account that P (1) = 600

$$p(1) = \frac{2}{3}k(1) + 500 = 600 \quad \rightarrow \quad \frac{2}{3}k = 100 \rightarrow k = 150$$

$$p(t) = \frac{2}{3}(150)t^{\frac{3}{2}} + 500 \rightarrow \boxed{p(t) = 100\sqrt{t^3} + 500}$$

The size of the population after 7 days is $\boxed{p(7) = 2352\ \textit{bacterias}.}$

A hot air balloon, which rises vertically at a speed of **16 feet** per second, drops a bag of sand at the instant it is **64 feet** above the ground.

013

a) How many seconds will the bag reach the ground? b) At what speed will it make contact with the ground?

Consider that t = 0 represents the initial time. The conditions indicated in the exercise can be written as follows

$$s(0) = 64 \; pies, s'(t) = 16\frac{pies}{s} \; y \; s''(t) = -32pies/s^2$$

a) To calculate in how many seconds the bag will reach the ground, the position function s (t) is required.

$$s'(t) = \int s''(t) \, dt = \int -32 \, dt = -32t + C_1$$

How speed is $16\frac{feet}{s}$, is obtained: $s'(0) = 16 = -32(0) + C_1$, This implies that $C_1 = 16$. Then you have

$$s'(t) = -32t + 16$$

According to s'(t) we get $s(t)$

$$s(t) = \int s'(t) \, dt = \int (-32t + 16) dt = -32 \int t \, dt + 16 \int dt$$

$$s(t) = -\frac{32}{2}t^2 + 16t + C_2 \;\; \rightarrow \;\; s(t) = -16t^2 + 16t + C_2$$

Taking the corresponding height you have

$$s(0) = -16(0)^2 + 16(0) + C_2 = 64 \Longrightarrow C_2 = 64$$

Consequently, you get:

$$s(t) = -16t^2 + 16t + 64$$

Equaling the previous equation to zero and clearing t

$$\boxed{t = 2,562 \ seconds. \ Time \ it \ takes \ to \ reach \ the \ ground}$$

b) As: $v(t) = s'(t) = -32t + 16.$ *and for* $t = 2,562$ *s is obtained*

$$\boxed{v(t) = -65,984 \ pies/s}$$

014

Above the moon, the acceleration of gravity is **-1.6 m/s²**. In the moon, a stone is dropped from a rock and hits the surface of the rock 20 seconds later. **a) From what height did it fall? b) What was your speed at the time of impact?**

Consider that t = 0 represents the initial time. The conditions indicated in the exercise can be written as follows: $v_0 = 0, \ S''(t) = -1.6 \ m/s^2$

a) To calculate at what height (s_0) the stone fell, the position function s (t) is required

$$s'(t) = v(t) = \int s''(t) \, dt \ = \int -1.6 \, dt = -1.6 \, t + v_0$$

$$s'(t) = -1.6 \, t$$

In function of $s'(t)$ is obtained $s(t)$

$$s(t) = \int s'(t) \, dt = \int -1.6 \, t \, dt = -\frac{1.6}{2} \, t^2 + s_0 = -0.8 \, t^2 + s_0$$

For t = 20 seconds, equaling the previous equation to zero and clearing S_O is obtained

$$s(20) = -0.8 \, (20)^2 + s_0 = 0 \rightarrow \boxed{s_0 = 320 \ metros}$$

b) For t = 20 seconds we have

$$v(20) = -1.6 \, t = -1.6(20) = -32 \ m/s \rightarrow \boxed{v(t) = -32 \ m/s}$$

Comment for the exercises 13 y 14

Demonstration of $s(t) = \int s'(t)\, dt$ y $s'(t) = \int s''(t)\, dt$

$$\frac{ds}{dt} = v(t) = s' \;\rightarrow\; ds = s'dt \;\rightarrow\; \int ds = \int s'dt \;\rightarrow\; s(t) = \int s'dt$$

$$\frac{dv}{dt} = a(t) = s'' \;\rightarrow\; dv = s''dt \;\rightarrow\; \int dv = \int s''dt \;\rightarrow\; v(t) = \int s''dt$$

As $v(t) = s'$ you have, $s'(t) = \int s''dt$

In the following exercises are the integrals applying the technique of substitution (change of variable)

$$\int (1+2x)^4 \, (2) dx$$

015

The substitution is made $u = 1 + 2x$ and it is derived

$$u = 1 + 2x \longrightarrow du = 2dx$$

As you can see the term $2dx$ that appears in the differential (du), is part of the original integral, so you can write

$$\int (1+2x)^4 \, (2) dx = \int u^4 \, du$$

Then we proceed to integrate in terms of u and the result in terms of x

$$\int (1+2x)^4 \, (2) dx = \frac{1}{5} u^5 + C = \frac{1}{5} \, (1+2x)^5 + C$$

$$\boxed{\int (1+2x)^4 \, (2) dx = \frac{1}{5}(1+2x)^5 + C}$$

$$\int \sqrt{9 - x^2} \, (-2x) dx$$

016

The substitution is made $u = 9 - x^2$ and it is derived

$$u = 9 - x^2 \longrightarrow du = -2xdx$$

Analogously to the previous exercise, we have

$$\int \sqrt{9 - x^2} \, (-2x) dx = \int \sqrt{u} \, du = \int u^{1/2} \, du = \frac{u^{3/2}}{3/2} + C$$

$$\boxed{\int \sqrt{9 - x^2} \, (-2x) dx = \frac{2}{3}(9 - x^2)^{3/2} + C}$$

017

$$\int \frac{1 + lnx}{3 + xlnx}\, dx$$

The following substitution is made

$$u = 3 + xlnx \;\longrightarrow\; du = \left(x\frac{1}{x} + lnx\right)dx \;\rightarrow\; du = (1 + lnx)dx$$

As you can see the term $(1 + lnx)\, dx$ that appears in the differential (du), is part of the original integral, so you can write

$$\int \frac{1 + lnx}{3 + xlnx}\, dx = \int \frac{du}{u} = \ln(u) + C = \ln(3 + xlnx) + C$$

$$\boxed{\int \frac{(1 + lnx)dx}{3 + xlnx} = \ln(3 + xlnx) + C}$$

018

$$\int \frac{dx}{x(lnx)^3}$$

The following substitution is made

$$u = lnx \;\longrightarrow\; du = \frac{dx}{x}$$

As you can see the term $\frac{dx}{x}$ that appears in the differential (du), is part of the original integral, so you can write

$$\int \frac{dx}{x(lnx)^3} = \int \frac{du}{u^3} = \int u^{-3}\, du = \frac{u^{-2}}{-2} + C = -\frac{1}{2(u)^2} + C$$

$$\boxed{\int \frac{dx}{x(lnx)^3} = -\frac{1}{2(lnx)^2} + C}$$

$$\int x\sqrt{x+2}\ dx$$

Let $u = x + 2$, $du = dx$. Since the integrand contains a factor x one has to clear x in terms of u, that is,

$$x = u - 2$$

Then the substitution is made and the integral is separated into two integrals

$$\int x\sqrt{x+2}\ dx = \int (u-2)\sqrt{u}\ du$$

$$\int x\sqrt{x+2}\ dx = \int u\sqrt{u}\ du - 2\int \sqrt{u}\ du$$

$$\int x\sqrt{x+2}\ dx = \int \sqrt{u^3}\ du - 2\int \sqrt{u}\ du$$

$$\int x\sqrt{x+2}\ dx = \int u^{\frac{3}{2}}\ du - 2\int u^{\frac{1}{2}}\ du$$

$$\int x\sqrt{x+2}\ dx = \frac{2}{5}u^{\frac{5}{2}} - \frac{4}{3}u^{\frac{3}{2}} + C = \frac{6\,U^{\frac{5}{2}} - 20\,U^{\frac{3}{2}}}{15} + C$$

Taking out common factor $\frac{2}{15}u^{\frac{3}{2}}$ you have

$$\int x\sqrt{x+2}\ dx = \frac{2}{15}u^{\frac{3}{2}}(3u-10) + C = \frac{2}{15}(x+2)^{\frac{3}{2}}[3(x+2)-10] + C$$

$$\int x\sqrt{x+2}\ dx = \frac{2}{15}(x+2)^{\frac{3}{2}}[3x+6-10] + C = \frac{2}{15}(x+2)^{\frac{3}{2}}[3x-4] + C$$

$$\boxed{\int x\sqrt{x+2}\ dx = \frac{2}{15}(x+2)^{\frac{3}{2}}[3x-4] + C}$$

020

$$\int \frac{e^t - 1}{e^t + 1}\, dt$$

From the original integral two integral

$$\int \frac{e^t - 1}{e^t + 1}\, dt = \int \frac{e^t dt}{e^t + 1} - \int \frac{1}{e^t + 1}\, dt$$

When making the substitution of u and du in the first integral its resolution is direct. Which does not happen for the second integral. To solve it, the following artifice must be applied. The integrand is multiplied and divided by e^{-t}, that is to say

$$\int \frac{dt}{e^t + 1} = \int \frac{e^{-t} dt}{e^{-t}(e^t + 1)} = \int \frac{e^{-t} dt}{1 + e^{-t}}$$

As a result, you get

$$\int \frac{e^t - 1}{e^t + 1}\, dt = \int \frac{e^t}{e^t + 1}\, dt - \int \frac{e^{-t} dt}{1 + e^{-t}}$$

Now we proceed to make the respective substitutions

$$u = e^t + 1 \; \rightarrow \; du = e^t dt; \; w = 1 + e^{-t} \; \rightarrow \; dw = -e^{-t} dt$$

Then,

$$\int \frac{e^t - 1}{e^t + 1}\, dt = \int \frac{e^t}{e^t + 1}\, dt + \int \frac{-e^{-t} dt}{1 + e^{-t}}$$

$$\int \frac{e^t - 1}{e^t + 1}\, dt = \int \frac{du}{u} + \int \frac{dw}{w} = \ln|u| + c_1 + \ln|w| + c_2$$

$$\int \frac{e^t - 1}{e^t + 1}\, dt = \ln(e^t + 1) + \ln(1 + e^{-t}) + C = \ln[(e^t + 1)(1 + e^{-t})] + C$$

$$\boxed{\int \frac{e^t - 1}{e^t + 1}\, dt = \ln[(e^t + 1)(1 + e^{-t})] + C}$$

$$\int \frac{e^{2x}\,dx}{\sqrt{e^x+1}}$$

Be $u = e^x + 1$, $du = e^x dx$. As you can see the factor e^x in the differential of (du), it does not correspond to the factor that is in the numerator of the integrand e^{2x}. Which makes it necessary to appley the following artifice: The factor e^{2x} It decomposes as the product of two powers: $e^{2x} = e^x.e^x$

Therefore, when rewriting the integral you have

$$\int \frac{e^{2x}\,dx}{\sqrt{e^x+1}} = \int \frac{e^x\,e^x\,dx}{\sqrt{e^x+1}}$$

Be $u = e^x + 1$, $du = e^x dx$. As the integrand still contains a factor e^x, it is cleared in terms of u, that is: $u = e^x + 1 \rightarrow e^x = u - 1$.

Now we proceed to make the corresponding substitutions

$$\int \frac{e^x\,e^x\,dx}{\sqrt{e^x+1}} = \int \frac{(u-1)\,du}{\sqrt{u}} = \int \frac{udu}{\sqrt{u}} - \int \frac{du}{\sqrt{u}}$$

$$\int \frac{e^x\,e^x\,dx}{\sqrt{e^x+1}} = \int u^{\frac{1}{2}}du - \int u^{-\frac{1}{2}}du$$

$$\int \frac{e^x\,e^x\,dx}{\sqrt{e^x+1}} = \frac{u^{\frac{3}{2}}}{3/2} + c_1 - \frac{u^{\frac{1}{2}}}{1/2} + c_2 = \frac{2}{3}u^{\frac{3}{2}} - 2u^{\frac{1}{2}} + C$$

$$\int \frac{e^{2x}\,dx}{\sqrt{e^x+1}} = \frac{2}{3}\sqrt{(e^x+1)^3} - 2\sqrt{e^x+1} + C$$

$$\boxed{\int \frac{e^{2x}\,dx}{\sqrt{e^x+1}} = \frac{2}{3}\sqrt{(e^x+1)^3} - 2\sqrt{e^x+1} + C}$$

022

$$\int \frac{e^x dx}{\sqrt{1 - e^{2x}}}$$

The integral is rewritten by applying the power property of a power

$$\int \frac{e^x dx}{\sqrt{1 - e^{2x}}} = \int \frac{e^x dx}{\sqrt{1 - (e^x)^2}}$$

The next substitution and its derivative is carried out

$$u = e^x \longrightarrow du = e^x dx$$

$$\int \frac{e^x dx}{\sqrt{1 - (e^x)^2}} = \int \frac{du}{\sqrt{1 - (u)^2}} = \int \frac{du}{\sqrt{1 - u^2}}$$

The resulting integral is immediate

$$\int \frac{du}{\sqrt{1 - u^2}} = arcsen\ u + C$$

As a result, you get

$$\boxed{\int \frac{e^x dx}{\sqrt{1 - e^{2x}}} = arc\ sen\ e^x + C}$$

023

$$\int \frac{dt}{3^t + 4}$$

If we do $u = 3^t + 4$, $du = 3^t lm3dt$, we observe that the factor appears 3^t in the differential that is not in the numerator of the integrand. Therefore, the respective substitution is not appropriate. The next mathematical artifice consists in constructing the denominator of the integrand in the numerator of the same. That is, multiply and divide the integral by 4. Then add and subtract 3^t.

$$\int \frac{dt}{3^t + 4} = \frac{1}{4} \int \frac{4dt}{3^t + 4} = \frac{1}{4} \int \frac{3^t + 4 - 3^t}{3^t + 4} dt$$

The integral is separated into two integrals

$$\frac{1}{4} \int \frac{3^t + 4 - 3^t}{3^t + 4} dt = \frac{1}{4} \int \frac{(3^t + 4)}{(3^t + 4)} dt - \frac{1}{4} \int \frac{3^t dt}{3^t + 4}$$

The resolution of the first integral is immediate. The second integral is made the next substitution

$$u = 3^t + 4 \quad \longrightarrow \quad du = 3^t ln3 dt$$

Note: The term ln3 in the differential (du) is multiplying and does not appear in the numerator of the integral, therefore, the integral between ln3 is divided so that the respective integral is not altered.

$$\int \frac{dt}{3^t + 4} = \frac{1}{4} \int dt - \frac{1}{4ln3} \int \frac{du}{u} = \frac{1}{4} t - \frac{1}{4ln3} ln|u| + C$$

$$\boxed{\int \frac{dt}{3^t + 4} = \frac{1}{4} t - \frac{ln|3^t + 4|}{4ln3} + C}$$

$$\boxed{\int \frac{6x^3 + x^2 - 2x + 1}{2x - 1} dx}$$ 　　　　024

When the degree of the dividend polynomial is greater than or equal to the degree of the divisor polynomial, it is necessary to previously divide polynomials.

The result of the division is

$$\frac{6x^3 + x^2 - 2x + 1}{2x - 1} = 3x^2 + 2x + \frac{1}{2x - 1}$$

Then we proceed to integrate each term of the result

$$\int \frac{6x^3 + x^2 - 2x + 1}{2x - 1}\, dx = 3\int x^2\, dx + 2\int x\, dx + \int \frac{dx}{2x - 1}$$

The first and second integral are immediate, the third integral is solved with the following substitution

$$u = 2x - 1 \;\longrightarrow\; du = 2dx$$

$$\int \frac{6x^3 + x^2 - 2x + 1}{2x - 1}\, dx = 3\int x^2\, dx + 2\int x\, dx + \frac{1}{2}\int \frac{du}{u}$$

$$\int \frac{6x^3 + x^2 - 2x + 1}{2x - 1}\, dx = x^3 + x^2 + \frac{1}{2}\ln|u| + C$$

$$\boxed{\int \frac{6x^3 + x^2 - 2x + 1}{2x - 1}\, dx = x^3 + x^2 + \frac{1}{2}\ln(2x - 1) + C}$$

025
$$\boxed{\int \frac{x^2 + 4x + 1}{x + 2}\, dx}$$

The exercise can be solved in two ways. The first one realizing the division of the polynomials (see previous problem). The second way is completing squares.

To complete squares the coefficient of x between two is divided and squared

$$(x^2 + 4x) + 1 = (x^2 + 4x + 4) + 1 - 4 = (x + 2)^2 - 3$$

Consequently, you have

$$\int \frac{x^2 + 4x + 1}{x + 2}\, dx = \int \frac{(x + 2)^2 - 3}{x + 2}\, dx = \int \frac{(x + 2)^2}{(x + 2)}\, dx - 3\int \frac{dx}{x + 2}$$

$$\int \frac{x^2 + 4x + 1}{x + 2}\, dx = \int (x + 2)\, dx - 3\int \frac{dx}{x + 2}$$

$$\int \frac{x^2 + 4x + 1}{x + 2} dx = \int x\, dx + 2 \int dx - 3 \int \frac{dx}{x + 2}$$

The first integral and the second are immediate. The third integral is made the next substitution $u = x + 2 \longrightarrow du = dx$

$$\boxed{\int \frac{x^2 + 4x + 1}{x + 2} dx = \frac{1}{2}x^2 + 2x - 3\ln(x + 2) + C}$$

$$\int \frac{2t - 10}{2t^2 - 3} dt$$

026

The integral is rewritten by factoring in the numerator, and then separated into two integrals.

$$\int \frac{2t - 10}{2t^2 - 3} dt = 2 \int \frac{t - 5}{2t^2 - 3} dt = 2 \int \frac{t\,dt}{2t^2 - 3} - 10 \int \frac{dt}{2t^2 - 3}$$

By direct substitution, the first integral is resolved, not the second. For the resolution it is required to transform its denominator in a way $u^2 - a^2$. Which leads to an immediate integral. See summary of integration tables.

$$2t^2 - 3 = \left(\sqrt{2}\,t\right)^2 - (\sqrt{3})^2$$

Consequently, you have

$$\int \frac{2t - 10}{2t^2 - 3} dt = 2 \int \frac{t\,dt}{2t^2 - 3} - 10 \int \frac{dt}{\left(\sqrt{2}\,t\right)^2 - \left(\sqrt{3}\right)^2}$$

Then substitutions and their derivatives are made

$$u = 2t^2 - 3 \longrightarrow du = 4t\,dt$$

$$w = \sqrt{2}\,t \longrightarrow dw = \sqrt{2}\,dt$$

$$\int \frac{2t - 10}{2t^2 - 3} \, dt = \frac{1}{2} \int \frac{du}{u} - \frac{10}{\sqrt{2}} \int \frac{dw}{w^2 - (\sqrt{3})^2}$$

$$\int \frac{2t - 10}{2t^2 - 3} \, dt = \frac{1}{2} \ln|2t^2 - 3| - \frac{10}{\sqrt{2}} \frac{1}{2\sqrt{3}} \ln \left| \frac{t\sqrt{2} - \sqrt{3}}{t\sqrt{2} + \sqrt{3}} \right| + C$$

$$\int \frac{2t - 10}{2t^2 - 3} \, dt = \frac{1}{2} \ln|2t^2 - 3| - \frac{5\sqrt{6}}{6} \ln \left| \frac{t\sqrt{2} - \sqrt{3}}{t\sqrt{2} + \sqrt{3}} \right| + C$$

$$\boxed{\int \frac{2t - 10}{2t^2 - 3} \, dt = \frac{1}{2} \ln|2t^2 - 3| - \frac{5\sqrt{6}}{6} \ln \left| \frac{t\sqrt{2} - \sqrt{3}}{t\sqrt{2} + \sqrt{3}} \right| + C}$$

027

$$\int \sqrt{\frac{\ln(x + \sqrt{x^2 + 1})}{1 + x^2}} \, dx$$

The integral is rewritten by applying radiation properties

$$\int \sqrt{\frac{\ln(x + \sqrt{x^2 + 1})}{1 + x^2}} \, dx = \int \frac{\sqrt{\ln(x + \sqrt{x^2 + 1})}}{\sqrt{1 + x^2}} \, dx$$

$$u = \ln(x + \sqrt{x^2 + 1}) \quad , \quad du = \frac{dx}{\sqrt{x^2 + 1}}$$

Now we proceed to explain algebraically the origin of du

$$du = \left[\frac{1}{x + \sqrt{x^2 + 1}} \left(1 + \frac{2x}{2\sqrt{x^2 + 1}} \right) \right] dx = \left[\frac{1}{x + \sqrt{x^2 + 1}} \left(1 + \frac{x}{\sqrt{x^2 + 1}} \right) \right] dx$$

$$du = \left[\frac{1}{x + \sqrt{x^2 + 1}} \left(\frac{\sqrt{x^2 + 1} + x}{\sqrt{x^2 + 1}} \right) \right] dx = \frac{\sqrt{x^2 + 1} + x}{x\sqrt{x^2 + 1} + (x^2 + 1)} \, dx$$

Then multiply by the conjugate of the numerator

$$du = \frac{(\sqrt{x^2+1}+x)}{(x\sqrt{x^2+1})+(x^2+1)}\frac{(\sqrt{x^2+1}-x)}{(\sqrt{x^2+1}-x)}dx$$

$$du = \frac{x^2+1-x^2}{x(x^2+1)-x^2\sqrt{x^2+1}+(x^2+1)\sqrt{x^2+1}-x(1+x^2)}dx = \frac{dx}{\sqrt{x^2+1}}$$

$$\int \sqrt{\frac{ln(x+\sqrt{x^2+1})}{1+x^2}}\,dx = \int \sqrt{u}\,du = \int u^{\frac{1}{2}}\,du = \frac{u^{\frac{1}{2}+1}}{\frac{1}{2}+1} = \frac{u^{\frac{3}{2}}}{\frac{3}{2}} + C$$

$$\boxed{\int \sqrt{\frac{ln(x+\sqrt{x^2+1})}{1+x^2}}\,dx = \frac{2}{3}\sqrt{\left[ln(x+\sqrt{x^2+1})\right]^3} + C}$$

$$\int e^{sen^2 x}\,sen\,2x\,dx$$

028

Note that the argument of the sine function in the exponent of (e) is (x). And the argument of the other sine function is (2x). For there to be uniformity in the arguments, the following trigonometric identity is required, which allows us to solve the integral.

$$sen^2 x = \frac{1-\cos 2x}{2}$$

Therefore, you have

$$\int e^{sen^2 x}\,sen\,2x\,dx = \int e^{\frac{1-\cos 2x}{2}}\,sen\,2x\,dx$$

The following substitution is made

$$u = \frac{1-\cos 2x}{2} \rightarrow u = \frac{1}{2}(1-\cos 2x)$$

$$du = \frac{1}{2}[-(-sen\,2x)2] \rightarrow du = sen\,2x\,dx$$

As a result, you get

$$\int e^{sen^2 x} sen\,2x dx = \int e^{\frac{1-\cos 2x}{2}} sen\,2x dx = \int e^u du$$

Then we proceed to integrate

$$\int e^{sen^2 x} sen\,2x dx = e^u + C = e^{\frac{1-\cos 2x}{2}} + C$$

$$\boxed{\int e^{sen^2 x} sen\,2x dx = e^{sen^2 x} + C}$$

029

$$\int \frac{dx}{sen\,x\cos x}$$

To solve the integral, the following trigonometric identity is required

$$sen\,2x = 2sen\,x\cos x$$

Then it clears $sen\,x\cos x$

$$sen\,x\cos x = \frac{1}{2} sen\,2x$$

Therefore, you have

$$\int \frac{dx}{sen\,x\cos x} = \int \frac{dx}{\frac{1}{2} sen\,2x} = 2\int \frac{1}{sen\,2x} dx$$

Como $\dfrac{1}{sen\,2x} = \csc 2x$

$$\int \frac{dx}{sen\,x\cos x} = 2\int \csc 2x\,dx$$

The following substitution is made

$$u = 2x \longrightarrow du = 2dx$$

$$\int \frac{dx}{sen\, x cos\, x} = \frac{2}{2} \int cscu\, du = \ln|csc\, u - ctg\, u| + C$$

$$\boxed{\int \frac{dx}{sen\, x cos\, x} = \ln|csc\, 2x - ctg\, 2x| + C}$$

$$\int \sqrt{1 + 3cos^2 x}\ sen\, 2xdx$$

030

The reader should be careful to observe that the angles of both the cosine and the sine are not equal. For such equality to exist, the following trigonometric identity must be used

$$cos^2 x = \frac{1 + \cos 2x}{2}$$

Now we proceed to perform the respective substitution

$$\int \sqrt{1 + 3cos^2 x}\ sen\, 2xdx = \int \sqrt{1 + 3\left(\frac{1 + \cos 2x}{2}\right)}\ sen\, 2xdx$$

$$\int \sqrt{1 + 3cos^2 x}\ sen\, 2xdx = \int \sqrt{\frac{5 + 3\cos 2x}{2}}\ sen\, 2xdx$$

Then substitutions are determined

$$u = \frac{5 + 3\cos 2x}{2} \longrightarrow u = \frac{1}{2}(5 + 3\cos 2x)$$

$$du = \frac{1}{2}(-3sen2x * 2) \rightarrow du = -3sen\, 2xdx$$

$$\int \sqrt{1 + 3cos^2x} \ sen \ 2xdx = -\frac{1}{3} \int \sqrt{u} \ du = -\frac{1}{3} \int u^{\frac{1}{2}} \ du$$

$$\int \sqrt{1 + 3cos^2x} \ sen \ 2xdx = -\frac{1}{3} \frac{u^{\frac{3}{2}}}{\frac{3}{2}} + c = -\frac{2}{9} u^{\frac{3}{2}} + C$$

$$\boxed{\int \sqrt{1 + 3cos^2x} \ sen \ 2xdx = -\frac{2}{9} \sqrt{\left(\frac{5 + 3\cos 2x}{2}\right)^3} + C}$$

In the following exercises are the integrals applying the technique of integration by part

$$\int xe^x dx$$

The integral corresponds to the product of two functions (See theoretical basis). We will call (u) the potential function (x) since it appears first that the exponential (e^x) in ALPES. And (dv) the rest of the integrand.

That is to say,

$$u = x \longrightarrow du = dx$$

$$dv = e^x dx \longrightarrow \int dv = \int e^x \, dx$$

$$v = e^x$$

Therefore, the integration by parts produces

$$\int u \, dv = u\,v - \int v \, du$$

$$\int xe^x dx = x\,e^x - \int e^x \, dx = xe^x - e^x + C$$

As e^x it's common factor you have

$$\boxed{\int xe^x dx = e^x(x - 1) + C}$$

032

$$\int x^2 \ln x \, dx$$

We will call (u) al (lnx) since it appears first in ALPES that the potential function x^2. And the rest of the integrand equal to (dv).

That is to say

$$u = \ln x \ \longrightarrow du = \frac{1}{x} dx$$

$$dv = x^2 dx \ \longrightarrow \int dv = \int x^2 \, dx$$

$$v = \frac{x^3}{3}$$

Therefore, the integration by parts produces

$$\int u \, dv = u \, v - \int v \, du$$

$$\int x^2 \ln x \, dx = \frac{x^3}{3} \ln x - \int \left(\frac{x^3}{3}\right)\left(\frac{1}{x}\right) dx$$

$$\int x^2 \ln x \, dx = \frac{x^3}{3} \ln x - \frac{1}{3}\int x^2 dx$$

$$\int x^2 \ln x \, dx = \frac{x^3}{3} \ln x - \frac{1}{3}\frac{x^3}{3} + C$$

As $\frac{x^3}{3}$ it's common factor you have

$$\boxed{\int x^2 \ln x \, dx = \frac{x^3}{3}\left(\ln x - \frac{1}{3}\right) + C}$$

$$\int arcsenx \; dx$$

Since the integrand has a single factor, it becomes (u) equal to (arcsenx) and (dv) equal to (dx).

$$u = arcsenx \longrightarrow du = \frac{1}{\sqrt{1 - x^2}} dx$$

$$dv = dx \longrightarrow \int dv = \int dx \longrightarrow v = x$$

Therefore, the integration by parts produces

$$\int u \; dv = u \, v - \int v \; du$$

$$\int arcsenx \; dx = arcsenx \, (x) - \int \frac{x}{\sqrt{1 - x^2}} dx$$

By solving the second integral by substitution you get

$$w = 1 - x^2 \longrightarrow dw = -2x$$

Consequently, we have

$$\int arcsenx \; dx = arcsenx \, (x) + \frac{1}{2} \int w^{-\frac{1}{2}} dw$$

$$\int arcsenx \; dx = arcsenx(x) + \frac{1}{2} \frac{w^{\frac{1}{2}}}{\frac{1}{2}} + C$$

$$\boxed{\int arcsenx \; dx = x \; arcsenx + \sqrt{1 - x^2} + C}$$

034

$$\int x^2 senx \, dx$$

We will call (u) the power function (p) x^2 since it appears first in ALPES that the function (S) sinx. And the rest of the integrand equal to dv.

$$u = x^2 \longrightarrow du = 2xdx$$

$$dv = senx \, dx \longrightarrow \int dv = \int senx \, dx \rightarrow v = -\cos x$$

Therefore, the integration by parts produces

$$\int x^2 senx \, dx = -x^2 \cos x - \int -\cos x(2x)dx$$

$$\int x^2 senx \, dx = -x^2 \cos x + \int \cos x(2x)dx \quad \textbf{\textit{Equation}} \ \textbf{1}$$

Integration by parts partially simplified the original integral. That means, that it is necessary to evaluate the integral that appears in the result $(\int 2x \cos xdx)$. Now according to ALPES (u) it is equal to $(2x)$, and the rest of the integrand is equal to dv.

$$u = 2x \longrightarrow du = 2dx$$

$$dv = cox \, dx \longrightarrow \int dv = \int cosx \, dx \rightarrow v = sen \, x$$

Therefore, the integration by parts produces

$$\int 2x \cos xdx = 2xsen \, x - 2 \int sen \, xdx = 2xsen \, x + 2cosx + C \quad \textbf{\textit{Equation}} \ \textbf{2}$$

Substituting the result of equation 2 in equation 1 is obtained

$$\boxed{\int x^2 sen \, xdx = -x^2 \cos x + 2xsen \, x + 2cosx + C}$$

$$\int sec^3 x dx$$

035

The integral has a single factor. If we apply ALPES: $u = sec^3x \; y \; dv = dx$. When deriving (u) we have: $du = 3sec^2 x sec\, x tan\, x dx$. It can be noted that the derivative of (u) is very complex. This means that the application of the rule is not convenient. To solve the integral it is rewritten in the following way

$$\int sec^3 x dx = \int sec\, x \; sec^2 x \; dx$$

In this case the selection of (u) and (dv) is at the discretion of the reader. However, the easiest factor to derive and the easiest to integrate is selected.

$$u = secx \;\longrightarrow\; du = secx\; tanx\; dx$$

$$dv = sec^2 \, dx \;\longrightarrow\; \int dv = \int sec^2 \, dx \;\to\; v = tanx$$

$$\int sec^3 x dx = sec\, x \tan x - \int sec\, x \tan x \tan x \; dx$$

$$\int sec^3 x dx = sec\, x \tan x - \int sec\, x \; tan^2 x \; dx$$

$$\int sec^3 x dx = sec\, x \tan x - \int sec\, x \; (sec^2 x - 1)dx = sec\, x \tan x - \int sec^3 x \; dx + \int sec\, x \; dx$$

As you can see on the right side of equality, the original integral appears.
This is grouped on the left side of equality and then cleared

$$\int sec^3 x dx + \int sec^3 x \; dx = sec\, x \tan x + \int sec\, x \; dx$$

$$2\int sec^3 x dx = sec\, x \tan x + \ln(secx + tanx) + C$$

$$\boxed{\int sec^3 x \; dx = \frac{1}{2}[sec\, x \tan x + \ln(secx + tanx)] + C}$$

036

$$\int x^2 \arctan x \, dx$$

We will call $u = \arctan x$ y $dv = x^2$. According to the order of appearance in ALPES.

$$u = arctanx \longrightarrow du = \frac{dx}{1+x^2}$$

$$dv = x^2 \, dx \longrightarrow \int dv = \int x^2 \, dx \rightarrow v = \frac{x^3}{3}$$

$$\int x^2 \arctan x \, dx = \frac{x^3}{3} \arctan x - \frac{1}{3}\int \frac{x^3}{1+x^2} dx \quad \textbf{Equation 1}$$

Now we proceed to solve the integral $\int \frac{x^3}{1+x^2} dx$

$$\int \frac{x^3}{1+x^2} dx = \int \left(x - \frac{x}{1+x^2}\right) dx = \int x dx - \int \frac{x}{1+x^2} dx$$

We will call

$$w = 1 + x^2 \longrightarrow dw = 2xdx$$

$$\int \frac{x^3}{1+x^2} dx = \int x dx - \int \frac{x}{1+x^2} dx = \int x dx - \frac{1}{2}\int \frac{dw}{w}$$

$$\int \frac{x^3}{1+x^2} dx = \frac{1}{2}x^2 - \frac{1}{2}ln|1+x^2| + C$$

$$\int \frac{x^3}{1+x^2} dx = \frac{1}{2}(x^2 - ln|1+x^2|) + C \quad \textbf{Equation 2}$$

Substitute the result of equation 2 in equation 1

$$\int x^2 \arctan x \, dx = \frac{x^3}{3} \arctan x - \frac{1}{3}\left[\frac{1}{2}(x^2 - ln|1+x^2|)\right] + C$$

$$\int x^2 \arctan x \, dx = \frac{1}{3}\left[x^3 \arctan x - \frac{1}{2}(x^2 - \ln(1 + x^2)) \right] + C$$

$$\int x^2 e^{-3x} dx \qquad\qquad 037$$

We will call: $u = x^2$ y $dv = e^{-3x} dx$. According to the order of appearance in ALPES.

$$u = x^2 \longrightarrow du = 2x dx$$

$$dv = e^{-3x} dx \longrightarrow \int dv = \int e^{-3x} dx \;\rightarrow\; v = -\frac{1}{3} e^{-3x}$$

Therefore, the integration by parts produces

$$\int x^2 e^{-3x} dx = -\frac{1}{3} x^2 e^{-3x} + \frac{2}{3} \int x \, e^{-3x} dx \quad \textbf{Equation 1}$$

Now we proceed to solve the integral $\int x \, e^{-3x} dx$

$$u = x \longrightarrow du = dx$$

$$dv = e^{-3x} dx \longrightarrow \int dv = \int e^{-3x} dx \;\rightarrow\; v = -\frac{1}{3} e^{-3x}$$

Therefore, the integration by parts produces

$$\int x^2 e^{-3x} dx = -\frac{1}{3} x \, e^{-3x} + \frac{1}{3} \int e^{-3x} dx$$

$$\int x^2 e^{-3x} dx = -\frac{1}{3} x \, e^{-3x} - \frac{1}{9} e^{-3x} + C \quad \textbf{Equation 2}$$

Substitute the result of equation 2 in equation 1

$$\int x^2 e^{-3x} dx = -\frac{1}{3} x^2 e^{-3x} + \frac{2}{3}\left(-\frac{1}{3} x \, e^{-3x} - \frac{1}{9} e^{-3x} \right) + C$$

$$\int x^2 e^{-3x}dx = -\frac{1}{3}x^2 e^{-3x} - \frac{2}{9}x\,e^{-3x} - \frac{2}{27}e^{-3x} + C$$

$$\boxed{\int x^2 e^{-3x}dx = -\frac{1}{3}e^{-3x}\left(x^2 + \frac{2}{3}x + \frac{2}{9}\right) + C}$$

038 $\qquad\boxed{\int x^3 e^{-x^2}dx}$

Before directly applying the technique of integration by parts, it is necessary to analyze the integral proposal. If we do $dv = e^{-x^2}dx$ it is difficult for us to find v. It is then necessary to rewrite the integral (case studied in the integration by change of variable). And this rewriting allows us to correctly apply the technique of integration by parts.

$$\int x^3 e^{-x^2}dx = \int x^2 e^{-x^2}xdx$$

$$w = -x^2 \;\longrightarrow\; dw = -2xdx,\;\; x^2 = -w$$

$$\int x^3 e^{-x^2}dx = \int x^2 e^{-x^2}xdx = -\frac{1}{2}\int -w\,e^w dw = \frac{1}{2}\int w\,e^w dw$$

We will call: $u = w$ y $dv = e^w dw$. According to the order of appearance in ALPES.

$$u = w \;\longrightarrow\; du = dw$$
$$dv = e^w dw \;\longrightarrow\; \int dv = \int e^w dw \;\rightarrow\; v = e^w$$

Therefore, the integration by parts produces

$$\int x^3 e^{-x^2}dx = \frac{1}{2}\int w\,e^w dw = \frac{1}{2}w\,e^w - \frac{1}{2}\int e^w dw$$

$$\int x^3 e^{-x^2}dx = \frac{1}{2}w\,e^w - \frac{1}{2}e^w + c = -\frac{1}{2}x^2 e^{-x^2} - \frac{1}{2}e^{-x^2} + c$$

$$\int x^3 e^{-x^2} dx = -\frac{1}{2} e^{-x^2}(x^2 + 1) + c$$

$$\boxed{\int x^3 e^{-x^2} dx = -\frac{1}{2} e^{-x^2}(x^2 + 1) + c}$$

$$\int \frac{x^2 dx}{\sqrt{1 + x}}$$

039

The integral is rewritten so that the product of two functions appears.

$$\int \frac{x^2 dx}{\sqrt{1 + x}} = \int x^2 (1 + x)^{-\frac{1}{2}} dx$$

The application of the rule is not appropriate. The selection of u and dv remains on the reader's part. The suggestion we give is to choose the easiest factor to derive and the easiest factor to integrate.

$$u = x^2 \longrightarrow du = 2x dx$$

$$dv = (1 + x)^{-\frac{1}{2}} dx \longrightarrow \int dv = \int (1 + x)^{-\frac{1}{2}} dx \rightarrow v = 2(1 + x)^{\frac{1}{2}}$$

Therefore, the integration by parts produces

$$\int x^2 (1 + x)^{-\frac{1}{2}} dx = 2x^2 (1 + x)^{\frac{1}{2}} - 4 \int x (1 + x)^{\frac{1}{2}} dx \rightarrow \textbf{Equation 1}$$

Now we proceed to solve the integral $\int x (1 + x)^{\frac{1}{2}} dx$

$$u = x \longrightarrow du = dx$$

$$dv = (1 + x)^{\frac{1}{2}} dx \longrightarrow \int dv = \int (1 + x)^{\frac{1}{2}} dx \rightarrow v = \frac{2}{3}(1 + x)^{\frac{3}{2}}$$

Therefore, the integration by parts produces

71

Therefore, the integration by parts produces

$$\int x\,(1+x)^{\frac{1}{2}}\,dx = \frac{2}{3}x(1+x)^{\frac{3}{2}} - \frac{2}{3}\int (1+x)^{\frac{3}{2}}\,dx$$

$$\int x\,(1+x)^{\frac{1}{2}}\,dx = \frac{2}{3}x(1+x)^{\frac{3}{2}} - \frac{2}{3}\frac{(1+x)^{\frac{5}{2}}}{\frac{5}{2}} + C$$

$$\int x\,(1+x)^{\frac{1}{2}}\,dx = \frac{2}{3}x(1+x)^{\frac{3}{2}} - \frac{4}{15}(1+x)^{\frac{5}{2}} + C \rightarrow \textbf{\textit{Equation 2}}$$

Substitute the result of equation 2 in equation 1

$$\int \frac{x^2\,dx}{\sqrt{1+x}} = 2x^2(1+x)^{\frac{1}{2}} - 4\left[\frac{2}{3}x(1+x)^{\frac{3}{2}} - \frac{4}{15}(1+x)^{\frac{5}{2}}\right] + C$$

$$\int \frac{x^2\,dx}{\sqrt{1+x}} = 2x^2(1+x)^{\frac{1}{2}} - \frac{8}{3}x(1+x)^{\frac{3}{2}} + \frac{16}{15}(1+x)^{\frac{5}{2}} + C$$

$$\int \frac{x^2\,dx}{\sqrt{1+x}} = 2x^2\sqrt{1+x} - \frac{8}{3}x\sqrt{(1+x)^3} + \frac{16}{15}\sqrt{(1+x)^5} + C$$

$$\int \frac{x^2\,dx}{\sqrt{1+x}} = \frac{2\sqrt{1+x}}{15}[15x^2 - 20x - 20x^2 + 8(1+x)^2] + C$$

$$\int \frac{x^2\,dx}{\sqrt{1+x}} = \frac{2\sqrt{1+x}}{15}[-5x^2 - 20x + 8x^2 + 16x + 8] + C$$

$$\int \frac{x^2\,dx}{\sqrt{1+x}} = \frac{2\sqrt{1+x}}{15}[3x^2 - 4x + 8] + C$$

$$\boxed{\int \frac{x^2\,dx}{\sqrt{1+x}} = \frac{2\sqrt{1+x}}{15}(3x^2 - 4x + 8) + C}$$

$$\int x^2 \cos x \; dx$$

040

We will call: $u = x^2$ y $dv = \cos x \; dx$. According to the order of appearance in ALPES.

$$u = x^2 \longrightarrow du = 2xdx$$

$$dv = \cos x \; dx \longrightarrow \int dv = \int \cos x \; dx \;\; \rightarrow \;\; v = sen x$$

Therefore, the integration by parts produces

$$\int x^2 \cos x \; dx = x^2 sen\, x - 2 \int x \, sen\, x \; dx \;\; \rightarrow \;\; \textbf{\textit{Equation 1}}$$

Now we proceed to solve the integral $\int x \, sen \, x \; dx$

$$u = x \longrightarrow du = dx$$

$$dv = sen x \; dx \longrightarrow \int dv = \int sen x \; dx \;\; \rightarrow \;\; v = -\cos x$$

Therefore, the integration by parts produces

$$\int x \, sen \, x \; dx = -x \cos x + \int \cos x \; dx$$

$$\int x \, sen \, x \; dx = -x \cos x + sen \, x + c \;\; \rightarrow \;\; \textbf{\textit{Equation 2}}$$

Substitute the result of equation 2 in equation 1

$$\int x^2 \cos x \; dx = x^2 sen\, x - 2[-x \cos x + sen \, x] + C$$

$$\boxed{\int x^2 \cos x \; dx = x^2 sen\, x + 2x \cos x - 2 \, sen\, x + C}$$

041

$$\int \frac{ln2x}{x^2}\,dx$$

The integral is rewritten as follows

$$\int \frac{ln2x}{x^2}\,dx = \int ln2x\, x^{-2}\,dx = \int x^{-2}ln2x\,dx$$

We will call $u = \text{ln}2x$ y $dv = x^{-2}dx$. According to the order of appearance in ALPES.

$$u = ln2x \longrightarrow du = \frac{dx}{2x}(2) = \frac{dx}{x}$$
$$dv = x^{-2}dx \longrightarrow \int dv = \int x^{-2}\,dx \rightarrow v = -\frac{1}{x}$$

Therefore, the integration by parts produces

$$\int x^{-2}ln2x\,dx = -\frac{1}{x}(ln2x) + \int \frac{dx}{x^2}$$

$$\int x^{-2}ln2x\,dx = -\frac{1}{x}(ln2x) + \int x^{-2}\,dx$$

$$\int x^{-2}ln2x\,dx = -\frac{1}{x}(ln2x) - \frac{1}{x} + C$$

$$\int x^{-2}ln2x\,dx = -\frac{1}{x}(\ln(2x) + 1) + C$$

$$\boxed{\int \frac{ln2x}{x^2}\,dx = -\frac{ln2x + 1}{x} + C}$$

$$\int \frac{x^3 e^{x^2}}{(x^2 + 1)^2} dx$$

042

The integral is rewritten as follows

$$\int \frac{x^3 e^{x^2}}{(x^2 + 1)^2} dx = \int \frac{x \; x^2 \; e^{x^2}}{(x^2 + 1)^2} dx$$

Explanation: Separates x^3 as the product of two powers $x \; x^2$. In order to remain in the integrand the factor $\frac{x \, dx}{(x^2+1)^2}$. And this factor becomes equal to dv. That is easy to integrate. And the factor $x^2 \, e^{x^2}$ igual a u. Which is easy to derive.

$$u = x^2 \, e^{x^2} \; \longrightarrow \; du = 2xe^{x^2}(x^2 + 1)dx$$
$$dv = \frac{x \, dx}{(x^2 + 1)^2} \; \longrightarrow \; \int dv = \int \frac{x \, dx}{(x^2 + 1)^2}$$

The integrals are solved

$$\int dv = \int \frac{x \, dx}{(x^2 + 1)^2}$$

$$\int dv = \int \frac{x \, dx}{(x^2 + 1)^2} \; \rightarrow \; v = \int x(x^2 + 1)^{-2} dx$$

be $w = (x^2 + 1) \; \rightarrow \; dw = 2xdx$

$$v = \frac{1}{2} \int w^{-2} \, dw \; \rightarrow \; v = -\frac{1}{2}(x^2 + 1)^{-1} \; \rightarrow \; v = -\frac{1}{2(x^2 + 1)}$$

Therefore, the integration by parts produces

$$\int \frac{x^3 e^{x^2}}{(x^2 + 1)^2} dx = x^2 \, e^{x^2} \left(-\frac{1}{2(x^2 + 1)} \right) + \int \frac{2xe^{x^2}(x^2 + 1)dx}{2(x^2 + 1)}$$

$$\int \frac{x^3 e^{x^2}}{(x^2+1)^2}\,dx = -\frac{x^2\,e^{x^2}}{2(x^2+1)} + \int x\,e^{x^2}\,dx$$

be

$$w = x^2 \longrightarrow dw = 2x\,dx$$

$$\int \frac{x^3 e^{x^2}}{(x^2+1)^2}\,dx = -\frac{x^2\,e^{x^2}}{2(x^2+1)} + \frac{1}{2}\int e^w\,dw$$

$$\int \frac{x^3 e^{x^2}}{(x^2+1)^2}\,dx = -\frac{x^2\,e^{x^2}}{2(x^2+1)} + \frac{1}{2}e^{x^2} + C$$

$$\int \frac{x^3 e^{x^2}}{(x^2+1)^2}\,dx = \frac{-x^2\,e^{x^2} + e^{x^2}(x^2+1)}{2(x^2+1)} + C$$

$$\int \frac{x^3 e^{x^2}}{(x^2+1)^2}\,dx = \frac{-x^2\,e^{x^2} + x^2 e^{x^2} + e^{x^2}}{2(x^2+1)} + C$$

$$\int \frac{x^3 e^{x^2}}{(x^2+1)^2}\,dx = \frac{e^{x^2}}{2(x^2+1)} + C$$

$$\boxed{\int \frac{x^3 e^{x^2}}{(x^2+1)^2}\,dx = \frac{e^{x^2}}{2(x^2+1)} + C}$$

043 $\quad\boxed{\displaystyle\int x\,sen^2 x\,dx}$

We will call u equal to x and dv equal to $sen^2 x\,dx$. According to the order of appearance in ALPES.

$$u = x \longrightarrow du = dx$$

$$dv = sen^2 x\,dx \longrightarrow \int dv = \int sen^2 x\,dx \;\rightarrow\; v = \int sen^2 x\,dx$$

To solve the integral $\int sen^2x dx$ we express the sen^2x for the following trigonometric identity $sen^2x = \frac{1-\cos 2x}{2}$.

$$v = \int sen^2x dx \quad \rightarrow \quad v = \int \frac{1 - \cos 2x}{2} dx$$

$$v = \frac{1}{2} \int dx - \frac{1}{2} \int \cos 2x dx \quad \rightarrow \quad v = \frac{1}{2}x - \frac{1}{4}sen2x$$

Therefore, the integration by parts produces

$$\int xsen^2x dx = x \left(\frac{1}{2}x - \frac{1}{4}sen2x\right) - \int \left(\frac{1}{2}x - \frac{1}{4}sen2x\right) dx$$

$$\int xsen^2x dx = \frac{x^2}{2} - \frac{x}{4}sen\,2x - \frac{1}{2}\int x\,dx + \frac{1}{4}\int sen\,2x\,dx$$

$$\int xsen^2x dx = \frac{x^2}{2} - \frac{x}{4}sen2x - \frac{x^2}{4} - \frac{1}{8}\cos2x + C$$

$$\boxed{\int xsen^2x dx = \frac{x^2}{4} - \frac{x}{4}sen\,2x - \frac{1}{8}\cos 2x + C}$$

$$\boxed{\int tan^2x\,sec^3x}$$

The integrand is expressed as a secant function using the following trigonometric identity:
$tan^2x = sec^2x - 1$

$$\int tan^2x\,sec^3x\,dx = \int (sec^2x - 1)sec^3x\,dx = \int sec^5x dx - \int sec^3x\,dx$$

$$\int tan^2x\,sec^3x\,dx = \int sec^5x dx - \int sec^3x\,dx \quad \textbf{\textit{Equation 1}}$$

The integral is solved $\int sec^5x dx$ and it is rewritten in the following way

$$\int sec^5x\, dx = \int sec^3x\, sec^2dx$$

$$u = sec^3x \longrightarrow du = 3sec^2x\, sec\, x \tan x\, dx \quad \rightarrow \quad du = 3sec^3x \tan x\, dx$$

$$dv = sec^2dx \longrightarrow \int dv = \int sec^2dx \quad \rightarrow \quad v = tanx$$

$$\int sec^5x\, dx = sec^3x \tan x - 3\int sec^3x \tan^2 dx$$

This partial result is replaced in equation 1

$$\int \tan^2x\, sec^3x\, dx = sec^3x \tan x - 3\int \tan^2x\, sec^3x\, dx - \int sec^3x\, dx$$

On the right side of equality the original integral appears, it is grouped and you ge

$$4\int \tan^2x\, sec^3x\, dx = sec^3x \tan x - \int sec^3x\, dx \quad \textbf{\textit{Equation 2}}$$

The integral $\int sec^3x\, dx$ is resolved in this book and is the number 035. Its result is replaced in equation 2. And it clears $\int \tan^2x\, sec^3x\, dx$

$$\boxed{\int \tan^2x\, sec^3x\, dx = \frac{1}{4}sec^3x \tan x - \frac{1}{8}[\, sec\, x \tan x + \ln(secx + tanx)] + C}$$

045 $\qquad \boxed{\int e^{\sqrt{x}}dx}$

We will call $u = e^{\sqrt{x}}$ and $dv = dx$

$$u = e^{\sqrt{x}} \rightarrow du = e^{\sqrt{x}}\frac{1}{2\sqrt{x}}dx \rightarrow du = \frac{e^{\sqrt{x}}dx}{2\sqrt{x}}$$

$$dv = dx \rightarrow \int dv = \int dx \rightarrow v = x$$

Therefore, the integration by parts produces

$$\int e^{\sqrt{x}}dx = x\,e^{\sqrt{x}} - \int \frac{x\,e^{\sqrt{x}}dx}{2\sqrt{x}} \quad \textbf{\textit{Equation 1}}$$

The integral is solved $\int \frac{x\,e^{\sqrt{x}}dx}{2\sqrt{x}}$:

The integral is rewritten by making the following substitution

$$w = \sqrt{x} \rightarrow dw = \frac{dx}{2\sqrt{x}}, x = w^2$$

$$\int \frac{x\,e^{\sqrt{x}}dx}{2\sqrt{x}} = \frac{1}{2}\int w^2 e^w\,dw$$

We will call $u = w^2$ y $dv = e^w dw$. According to the order of appearance in ALPES.

$$u = w^2 \rightarrow du = 2wdw$$

$$dv = e^w dw \rightarrow \int dv = \int e^w dw \rightarrow v = e^w$$

Therefore, the integration by parts produces

$$\frac{1}{2}\int w^2 e^w\,dw = \frac{1}{2}[w^2\,e^w - 2\int w\,e^w dw] = \frac{1}{2}w^2\,e^w - \int w\,e^w dw$$

$$\int \frac{x\,e^{\sqrt{x}}dx}{2\sqrt{x}} = \frac{1}{2}w^2\,e^w - \int w\,e^w dw \quad \textbf{\textit{Equation 2}}$$

Substitute equation 2 in equation 1

$$\int e^{\sqrt{x}}dx = x\,e^{\sqrt{x}} - \frac{1}{2}w^2\,e^w + \int w\,e^w dw \quad \textbf{\textit{Equation 3}}$$

The integral is solved $\int w\,e^w dw$

We will call $u = w$ y $dv = e^w dw$. According to the order of appearance in ALPES.

$$u = w \rightarrow du = dw$$

$$dv = e^w dw \; \rightarrow \int dv = \int e^w dw \rightarrow v = e^w$$

Therefore, the integration by parts produces

$$\int w \; e^w dw = we^w - \int e^w dw = we^w - e^w + C \; \textbf{\textit{Equation 4}}$$

Substitute equation 4 in equation 3

$$\int e^{\sqrt{x}} dx = x \, e^{\sqrt{x}} - \frac{1}{2} w^2 \, e^w + we^w - e^w + C$$

$$\int e^{\sqrt{x}} dx = x \, e^{\sqrt{x}} - \frac{1}{2} x \, e^{\sqrt{x}} + \sqrt{x} \; e^{\sqrt{x}} - e^{\sqrt{x}} + C$$

$$\boxed{\int e^{\sqrt{x}} dx = \; e^{\sqrt{x}} \left[\frac{1}{2} x + \sqrt{x} - 1 \right] + C}$$

046 $\qquad\qquad\qquad\qquad\qquad\qquad\qquad\qquad\qquad \boxed{\int arctan\sqrt{x} \; dx}$

We will call $u = \; arctan\sqrt{x}, \; dv = dx$

$$du = \frac{dx}{1 + (\sqrt{x})^2} \frac{1}{2\sqrt{x}} \; \rightarrow \; du = \frac{dx}{1 + x} \frac{1}{2\sqrt{x}} \; \rightarrow \; du = \frac{\sqrt{x} dx}{2(1 + x)x}$$

$$dv = dx \; \rightarrow \int dv = \int dx \rightarrow v = x$$

Therefore, the integration by parts produces

$$\int arctan\sqrt{x} dx = x arctan\sqrt{x} - \frac{1}{2} \int \frac{x\sqrt{x} dx}{(1 + x)x}$$

$$\int arctan\sqrt{x} dx = x arctan\sqrt{x} - \frac{1}{2} \int \frac{\sqrt{x} dx}{1 + x} \qquad \textbf{\textit{Equation 1}}$$

The integral is solved $\int \frac{\sqrt{x}\,dx}{1+x}$:

The integral is rewritten with the following substitution $\sqrt{x} = t \quad x = t^2 \quad dx = 2t\,dt$

$$\int \frac{\sqrt{x}\,dx}{1+x} = \int \frac{t\,2t\,dt}{1+t^2} = 2\int \frac{t^2\,dt}{1+t^2}$$

This integral is substituted in equation 1

$$\int arctan\sqrt{x}\,dx = x\,arctan\sqrt{x} - \int \frac{t^2\,dt}{1+t^2}$$

To solve the integral, the division of the polynomials takes place

$$\int arctan\sqrt{x}\,dx = x\,arctan\sqrt{x} - \int \left(1 - \frac{1}{1+t^2}\right)dt = x\,arctan\sqrt{x} - \int dt + \int \frac{dt}{1+t^2}$$

$$\int arctan\sqrt{x}\,dx = x\,arctan\sqrt{x} - t + \arctan t + c$$

$$\boxed{\int \boldsymbol{arctan\sqrt{x}\ dx = x\,arctan\sqrt{x} - \sqrt{x} + \arctan\sqrt{x} + c}}$$

$$\boxed{\int \frac{sen^2 x\,dx}{e^x}} \qquad\qquad 047$$

$$\int \frac{sen^2 x\,dx}{e^x} = \int sen^2 x\ e^{-x}\,dx$$

$$u = sen^2 x \quad y \quad dv = e^{-x}dx$$

$$u = sen^2 x \rightarrow du = 2sen\,x\,\cos x\,dx$$

$$dv = e^{-x}dx \ \rightarrow \int dv = \int e^{-x}dx \rightarrow v = -e^{-x}$$

$$\int sen^2 x \, e^{-x} \, dx = -e^{-x} \, sen^2 x + \int e^{-x} \, 2sen \, x \cos x \, dx$$

$$\int sen^2 x \, e^{-x} \, dx = -e^{-x} \, sen^2 x + \int sen \, 2x \, e^{-x} \, dx \quad \textbf{Equation 1}$$

Integration by parts is applied to the integral: $\int sen \, 2x \, e^{-x} \, dx$

$$u = sen \, 2x \rightarrow du = 2 \cos 2x \, dx$$

$$dv = e^{-x} dx \rightarrow \int dv = \int e^{-x} dx \rightarrow v = -e^{-x}$$

$$\int sen \, 2x \, e^{-x} \, dx = -e^{-x} sen \, 2x + 2 \int e^{-x} \cos 2x \, dx \quad \textbf{Equation 2}$$

Substitute equation 2 in equation 1

$$\int sen^2 x \, e^{-x} \, dx = -e^{-x} \, sen^2 x - e^{-x} sen \, 2x + 2 \int e^{-x} \cos 2x \, dx \quad \textbf{Equation 3}$$

The integral is now resolved $\int e^{-x} \cos 2x \, dx$

$$u = \cos 2x \rightarrow du = -2 \, sen \, 2x \, dx$$

$$dv = e^{-x} dx \rightarrow \int dv = \int e^{-x} dx \rightarrow v = -e^{-x}$$

$$\int e^{-x} \cos 2x \, dx = -e^{-x} \cos 2x - 2 \int e^{-x} \, sen \, 2x \, dx \quad \textbf{Equation 4}$$

Substitute equation 4 in equation 3

$$\textbf{Equation 5}$$
$$\int sen^2 x \, e^{-x} \, dx = -e^{-x} \, sen^2 x - e^{-x} sen \, 2x - 2e^{-x} \cos 2x - 4 \int e^{-x} \, sen \, 2x \, dx$$

The integral is now resolved $\int e^{-x} \, sen \, 2x \, dx$

$$u = sen \, 2x \rightarrow du = 2 \cos 2x \, dx$$

$$dv = e^{-x} dx \rightarrow \int dv = \int e^{-x} dx \rightarrow v = -e^{-x}$$

$$\int e^{-x} \operatorname{sen} 2x \, dx = -e^{-x} \operatorname{sen} 2x + 2 \int e^{-x} \cos 2x \, dx \quad \textbf{\textit{Equation 6}}$$

Relating equation 6 with equation 4 we have

$$\int e^{-x} \operatorname{sen} 2x \, dx = -e^{-x} \operatorname{sen} 2x + 2 \left[-e^{-x} \cos 2x - 2 \int e^{-x} \operatorname{sen} 2x \, dx \right]$$

$$\int e^{-x} \operatorname{sen} 2x \, dx = -e^{-x} \operatorname{sen} 2x - 2e^{-x} \cos 2x - 4 \int e^{-x} \operatorname{sen} 2x \, dx$$

$$5 \int e^{-x} \operatorname{sen} 2x \, dx = -e^{-x} \operatorname{sen} 2x - 2e^{-x} \cos 2x$$

$$\int e^{-x} \operatorname{sen} 2x \, dx = -\frac{1}{5} e^{-x} \operatorname{sen} 2x - \frac{2}{5} e^{-x} \cos 2x \quad \textbf{\textit{Equation 7}}$$

Equation 7 is substituted in equation 5 and by operating we obtain

$$\int \operatorname{sen}^2 x \, e^{-x} \, dx = -e^{-x} \operatorname{sen}^2 x - \frac{1}{5} e^{-x} \operatorname{sen} 2x - \frac{2}{5} e^{-x} \cos 2x$$

$$\boxed{\int \frac{\operatorname{sen}^2 x \, dx}{e^x} = -e^{-x} \operatorname{sen}^2 x - \frac{1}{5} e^{-x} (\operatorname{sen} 2x + 2 \cos 2x) + c}$$

$$\int x \, 2^{-x} \, dx \qquad\qquad\qquad 048$$

You select the easiest factor to derive and the easiest to integrate. The x factor is easier to derive. So, $dv = 2^{-x} dx$. Keep in mind that the integral of a constant raised to a function is $\int a^u du = \frac{a^u}{\ln a} + c$.

$$u = x \;\rightarrow\; du = dx, dv = 2^{-x} dx \;\rightarrow\; \int dv = \int 2^{-x} dx \;\rightarrow\; v = -\frac{2^{-x}}{\ln 2}$$

Therefore, the integration by parts produces

$$\int x \, 2^{-x} \, dx = -\frac{x}{2^x \ln 2} + \int \frac{2^{-x}}{\ln 2} dx = -\frac{x}{2^x \ln 2} + \frac{1}{\ln 2} \int 2^{-x} dx$$

$$\int x\, 2^{-x}\, dx = -\frac{x}{2^x \ln 2} + \frac{1}{\ln 2}\left(-\frac{2^{-x}}{\ln 2}\right) + C$$

$$\int x\, 2^{-x}\, dx = -\frac{x}{2^x \ln 2} - \frac{1}{2^x \ln^2 2} + C$$

$$\boxed{\int x\, 2^{-x}\, dx = -\frac{x}{2^x \ln 2} - \frac{1}{2^x \ln^2 2} + C}$$

049

$$\boxed{\int \frac{x\, dx}{sen^2 x} = \int x\, csc^2 x\, dx}$$

$$u = x \;\rightarrow\; du = dx$$

$$dv = csc^2 x\, dx \rightarrow \int dv = \int csc^2 x\, dx \rightarrow v = -cotx$$

$$\int \frac{x\, dx}{sen^2 x} = -x\cot x + \int \cot x\, dx = -x\cot x + ln|sen\, x| + C$$

$$\boxed{\int \frac{x\, dx}{sen^2 x} = -x\cot x + ln|sen\, x| + C}$$

050

$$\boxed{\int e^{2x} sen\, x\, dx}$$

According to the order of appearance in ALPES, $u = e^{2x}$ y $dv = sen\, xdx$.

$$u = e^{2x} \rightarrow du = 2e^{2x}dx \quad dv = sen\, x\, dx \rightarrow \int dv = \int sen\, x\, dx \rightarrow v = -cosx$$

Therefore, the integration by parts produces

$$\int e^{2x} sen\, x\, dx = -e^{2x}\cos x + 2\int e^{2x}\cos x\, dx \quad \textbf{\textit{Equation 1}}$$

The integral is now resolved $\int e^{2x} \cos x \, dx$

$$u = e^{2x} \rightarrow du = 2e^{2x}dx \longrightarrow dv = \cos x \, dx \rightarrow \int dv = \int \cos x \, dx \rightarrow v = senx$$

Therefore, the integration by parts produces

$$\int e^{2x} \cos x \, dx = e^{2x}sen\,x - 2\int e^{2x} sen\,x dx + c \; \textbf{\textit{Equation 2}}$$

Substitute equation 2 in equation 1

$$\int e^{2x}sen\,x \, dx = -e^{2x}\cos x + 2\left[e^{2x}sen\,x - 2\int e^{2x}sen\,x dx\right] + C$$

$$\int e^{2x}sen\,x \, dx = -e^{2x}\cos x + 2\,e^{2x}sen\,x - 4\int e^{2x}sen\,x dx + C$$

Observe the reader, the appearance of the original integral in the previous result, that leads us to group it with the original integral in the first member of the equation, and then proceed to clear the integral.

$$\int e^{2x}sen\,x \, dx + 4\int e^{2x}sen\,x dx = -e^{2x}\cos x + 2\,e^{2x}sen\,x + C$$

$$5\int e^{2x}sen\,x \, dx = -e^{2x}\cos x + 2\,e^{2x}sen\,x + C$$

$$\boxed{\int e^{2x}sen\,x \, dx = \frac{1}{5}e^{2x}(2sen\,x - \cos x) + C}$$

051

$$\int e^x \cos 2x dx$$

According to the order of appearance in ALPES $u = e^x$ y $dv = cos2xdx$.

$$u = e^x \;\rightarrow\; du = e^x dx$$

$$dv = cos2\,x\,dx \rightarrow \int dv = \int cos\,2x\,dx \;\rightarrow\; v = \frac{1}{2}sen2x$$

Therefore, the integration by parts produces

$$\int e^x \cos 2x dx = \frac{1}{2}e^x \, \text{sen} 2\,x - \frac{1}{2}\int e^x \, \text{sen} 2\,x\,dx \quad \textbf{Equation 1}$$

The integral is now resolved $\int e^x \, \text{sen} 2\,x\,dx$:

$$u = e^x \;\rightarrow\; du = e^x dx$$

$$dv = sen2\,x\,dx \rightarrow \int dv = \int sen\,2x\,dx \;\rightarrow\; v = -\frac{1}{2}cos2x$$

Therefore, the integration by parts produces

$$\int e^x \, \text{sen} 2\,x\,dx = -\frac{1}{2}e^x \cos 2\,x + \frac{1}{2}\int e^x \cos 2\,x\,dx \quad \textbf{Equation 2}$$

Substitute equation 2 in equation 1

$$\int e^x \cos 2x dx = \frac{1}{2}e^x \, \text{sen} 2\,x - \frac{1}{2}\left[-\frac{1}{2}e^x \cos 2\,x + \frac{1}{2}\int e^x \cos 2\,x\,dx\right]$$

$$\int e^x \cos 2x dx = \frac{1}{2}e^x \, \text{sen} 2\,x + \frac{1}{4}e^x \cos 2\,x - \frac{1}{4}\int e^x \cos 2\,x\,dx$$

$$\int e^x \cos 2x dx + \frac{1}{4}\int e^x \cos 2\,x\,dx = \frac{1}{2}e^x \, \text{sen} 2\,x + \frac{1}{4}e^x \cos 2\,x + C$$

$$\frac{5}{4}\int e^x \cos 2x dx = \frac{1}{2}e^x \, \text{sen} 2\,x + \frac{1}{4}e^x \cos 2\,x + C$$

$$\boxed{\int e^x \cos 2x dx = \frac{1}{5}e^x(2\, \textbf{sen}\, 2\,x + \cos 2\,x) + C}$$

$$\int \frac{x}{\sqrt{2+3x}}dx$$

052

The integral is rewritten

$$\int \frac{x}{\sqrt{2+3x}}dx = \int x(2+3x)^{-\frac{1}{2}}dx$$

You select the easiest factor to derive and the easiest to integrate. The x factor is easier to derive. The factor dv is equal to $(2+3x)^{-\frac{1}{2}}$.

$$u = x \rightarrow du = dx$$

$$dv = (2+3x)^{-\frac{1}{2}}dx \rightarrow \int dv = \int (2+3x)^{-\frac{1}{2}}dx \rightarrow v = \frac{2}{3}\sqrt{2+3x}\,dx$$

Therefore, the integration by parts produces

$$\int \frac{x}{\sqrt{2+3x}}dx = \frac{2}{3}x\sqrt{2+3x} - \frac{2}{3}\int \sqrt{2+3x}\,dx$$

$$\int \frac{x}{\sqrt{2+3x}}dx = \frac{2}{3}x\sqrt{2+3x} - \frac{2}{3}\left[\frac{2}{9}(2+3x)^{\frac{3}{2}}\right] + C$$

$$\int \frac{x}{\sqrt{2+3x}}dx = \frac{2}{3}x\sqrt{2+3x} - \frac{4}{27}\sqrt{(2+3x)^3} + C$$

$$\int \frac{x}{\sqrt{2+3x}}dx = \frac{2\sqrt{2+3x}}{3}\left[x - \frac{2}{9}(2+3x)\right] + C$$

$$\int \frac{x}{\sqrt{2+3x}}dx = \frac{2\sqrt{2+3x}}{27}[9x - 2(2+3x)] + C$$

$$\boxed{\int \frac{x}{\sqrt{2+3x}}dx = \frac{2\sqrt{2+3x}}{27}(3x-4) + C}$$

053

$$\int \theta \sec \theta \tan \theta \, d\theta$$

Select θ and make it equal to u and dv equal to sec θ tan θ dθ.

$$u = \theta \rightarrow du = d\theta$$

$$dv = \sec\theta\,\tan\theta d\theta \rightarrow \int dv = \int \sec\theta\,\tan\theta\,d\theta \rightarrow v = \sec\theta$$

Therefore, the integration by parts produces

$$\int \theta\sec\theta\,\tan\theta\,d\theta = \theta\sec\theta - \int \sec\theta d\theta = \theta\sec\theta - ln|\sec\theta + \tan\theta| + C$$

$$\boxed{\int \theta \sec\theta \tan\theta \, d\theta = \theta \sec\theta - ln|\sec\theta + \tan\theta| + C}$$

054

$$\int \frac{2x}{e^x} dx$$

The integral is rewritten

$$\int \frac{2x}{e^x} dx = 2\int xe^{-x}dx$$

Select x and make it equal to u and dv equal to $e^{-x}dx$

$$u = x \rightarrow du = dx$$

$$dv = e^{-x}dx \rightarrow \int dv = \int e^{-x}dx \rightarrow v = -e^{-x}$$

$$2\int xe^{-x}dx = -xe^{-x} + \int e^{-x}dx \rightarrow \int x\,e^{-x}dx = -2xe^{-x} - 2e^{-x} + C$$

$$\boxed{\int \frac{2x}{e^x} dx = -2(xe^{-x} + e^{-x}) + C}$$

$$\int \frac{xe^{2x}}{(2x+1)^2}dx$$

055

It is selected xe^{2x} which represents the derivative of a product, and $dv = \frac{dx}{(2x+1)^2}$ el factor a integrar.

$$u = xe^{2x} \rightarrow du = (2xe^{2x} + e^{2x})dx$$
$$du = e^{2x}(2x+1)dx$$
$$dv = \frac{dx}{(2x+1)^2} \rightarrow \int dv = \int \frac{dx}{(2x+1)^2} \rightarrow v = -\frac{1}{2(2x+1)}$$

Therefore, the integration by parts produces

$$\int \frac{xe^{2x}}{(2x+1)^2}dx = -\frac{xe^{2x}}{2(2x+1)} + \int \frac{e^{2x}(2x+1)dx}{2(2x+1)}$$

$$\int \frac{xe^{2x}}{(2x+1)^2}dx = -\frac{xe^{2x}}{2(2x+1)} + \frac{1}{2}\int e^{2x}dx$$

$$\int \frac{xe^{2x}}{(2x+1)^2}dx = -\frac{xe^{2x}}{2(2x+1)} + \frac{1}{4}e^{2x} + C$$

$$\int \frac{xe^{2x}}{(2x+1)^2}dx = \frac{-2xe^{2x} + 2xe^{2x} + e^{2x}}{4(2x+1)} + C$$

$$\int \frac{xe^{2x}}{(2x+1)^2}dx = \frac{-2xe^{2x} + e^{2x}(2x+1)}{4(2x+1)} + C$$

$$\int \frac{xe^{2x}}{(2x+1)^2}dx = \frac{-2xe^{2x} + 2xe^{2x} + e^{2x}}{4(2x+1)} + C$$

$$\boxed{\int \frac{xe^{2x}}{(2x+1)^2}dx = \frac{e^{2x}}{4(2x+1)} + C}$$

056

$$\int sen\,x\cos 2x\,dx$$

It is selected $u = \cos 2x \; y \; dv = senx\,dx$

$$u = \cos 2x \;\rightarrow\; du = -2sen\,2xdx$$

$$dv = sen\,x\,dx \;\rightarrow\; \int dv = \int sen\,x\,dx \;\rightarrow\; v = -\cos x$$

Therefore, the integration by parts produces

$$\int sen\,x\cos 2x\,dx = -\cos x\cos 2x - 2\int sen\,2x\cos x\,dx \quad \textbf{\textit{Equation 1}}$$

The integral is solved $\int sen\,2x\cos x\,dx$

$$u = sen\,2x \;\rightarrow\; du = 2cos\,2xdx$$

$$dv = cos\,x\,dx \;\rightarrow\; \int dv = \int cos\,x\,dx \;\rightarrow\; v = senx$$

$$\int sen\,2x\cos x\,dx = sen\,x\,sen\,2x - 2\int sen\,x\cos 2x\,dx \quad \textbf{\textit{Equation 2}}$$

Substitute equation 2 in equation 1

$$\int sen\,x\cos 2x\,dx = -\cos x\cos 2x - 2\left[sen\,x\,sen\,2x - 2\int sen\,x\cos 2x\,dx\right]$$

$$\int sen\,x\cos 2x\,dx = -\cos x\cos 2x - 2sen\,x\,sen\,2x + 4\int sen\,x\cos 2x\,dx$$

$$\int sen\,x\cos 2x\,dx - 4\int sen\,x\cos 2x\,dx = -\cos x\cos 2x - 2sen\,x\,sen\,2x + C$$

$$-3\int sen\,x\cos 2x\,dx = -\cos x\cos 2x - 2sen\,x\,sen\,2x + C$$

$$\boxed{\int sen\,x\cos 2x\,dx = \frac{1}{3}\cos x\cos 2x + \frac{2}{3}sen\,x\,sen\,2x + C}$$

$$\int (t \ln t)^2 \, dt$$

The integral is rewritten $\displaystyle\int (t \ln t)^2 \, dt = \int t^2 \ln^2 t \, dt$

It is selected $u = ln^2 t \ \ y \ \ dv = t^2 dt$.

$$u = (\ln t)^2 \rightarrow du = \frac{2 \ln t}{t} dt$$

$$dv = t^2 dt \rightarrow \int dv = \int t^2 dt \ \rightarrow \ v = \frac{t^3}{3}$$

$$\int t^2 \ln^2 t \, dt = \frac{1}{3} t^3 \ln^2 t - \frac{2}{3} \int \frac{\ln t \ \ t^3 dt}{t}$$

$$\int (t \ln t)^2 \, dt = \frac{1}{3} t^3 \ln^2 t - \frac{2}{3} \int \ln t \ t^2 \, dt \quad \textit{Equation 1}$$

The integral is solved $\int \ln t \ \ t^2 dt$

$$u = lnt \rightarrow du = \frac{dt}{t} \longrightarrow dv = t^2 dt \ \rightarrow \ \int dv = \int t^2 dt \ \rightarrow \ v = \frac{t^3}{3}$$

$$\int \ln t \ t^2 dt = \frac{1}{3} t^3 lnt - \frac{1}{3} \int \frac{t^3 dt}{t} = \frac{1}{3} t^3 lnt - \frac{1}{3} \int t^2 dt$$

$$\int \ln t \ t^2 dt = \frac{1}{3} t^3 lnt - \frac{1}{9} t^3 + C \ \textit{Equation 2}$$

Substitute equation 2 in equation 1

$$\int (t \ln t)^2 \, dt \frac{1}{3} t^3 ln^2 t - \frac{2}{3} \left[\frac{1}{3} t^3 lnt - \frac{1}{9} t^3 \right] + C$$

$$\boxed{\int (t \ln t)^2 \, dt = \frac{1}{3} t^3 ln^2 t - \frac{2}{9} t^3 lnt + \frac{2}{27} t^3 + C}$$

058

$$\int x^2 \ln x^2 dx$$

The integral is rewritten by applying the following artifice: The factor x^2 It decomposes as the product of two powers, that is, $x * x$. Then proceed to make the following substitution: $w = x^2$. Establish the difference of the previous exercise with the current one!

$$\int x^2 \ln x^2 dx = \int x\, x \ln x^2 dx$$

$$w = x^2 \rightarrow x = \sqrt{w} \rightarrow dw = 2xdx \rightarrow dx = \frac{dw}{2x}$$

$$\int x\, x \ln x^2 dx = \frac{1}{2}\int \sqrt{w}\, x \ln w\, \frac{dw}{2x} = \frac{1}{4}\int w^{\frac{1}{2}} \ln w\, dw$$

$$u = \ln w \rightarrow du = \frac{dw}{w}$$

$$dv = w^{\frac{1}{2}} dw \rightarrow \int dv = \int w^{\frac{1}{2}} dw \rightarrow v = \frac{2w^{3/2}}{3}$$

Therefore, the integration by parts produces

$$\int w^{\frac{1}{2}} \ln w\, dw = \frac{1}{4}\left[\frac{2w^{\frac{3}{2}} \ln w}{3} - \frac{2}{3}\int \frac{w^{3/2}}{w} dw\right]$$

$$\int w^{\frac{1}{2}} \ln w\, dw = \frac{1}{4}\left[\frac{2w^{\frac{3}{2}} \ln w}{3} - \frac{2}{3}\int w^{\frac{1}{2}} dw\right] = \frac{2w^{\frac{3}{2}} \ln w}{3} - \frac{2}{3}\left(\frac{2}{3}w^{\frac{3}{2}}\right)\right] + C$$

$$\int x^2 \ln x^2 dx = \frac{1}{4}\left[\frac{2w^{\frac{3}{2}} \ln w}{3} - \frac{4}{9}w^{\frac{3}{2}}\right] + C$$

$$\boxed{\int x^2 \ln x^2 dx = \frac{1}{6}x^3 \ln x^2 - \frac{1}{9}x^3 + C}$$

$$\int x^5 e^{2x^3} dx$$

The integral is rewritten by applying the following artifice: The factor x^5 decomposes as the product of two powers, that is, $x^3 * x^2$. Then proceed to make the following substitution: $w = 2x^3$.

$$\int x^5 e^{2x^3} dx = \int x^3 x^2 e^{2x^3} dx$$

$$w = 2x^3 \longrightarrow dw = 6x^2 dx \longrightarrow dx = \frac{dw}{6x^2}, \quad x^3 = \frac{w}{2}$$

$$\int x^3 x^2 e^{2x^3} dx = \int \frac{w}{2} x^2 e^w \frac{dw}{6x^2} = \frac{1}{12}\int w \; e^w dw$$

Therefore, the integration by parts produces

$$u = w \rightarrow du = dw$$

$$dv = e^w dw \rightarrow \int dv = \int e^w dw \rightarrow v = e^w$$

Therefore, the integration by parts produces

$$\frac{1}{12}\int w \; e^w dw = \frac{1}{12}\left[we^w - \int e^w dw \right]$$

$$\int x^3 x^2 e^{2x^3} dx = \frac{1}{12}\left(2x^3 e^{2x^3} - e^{2x^3} \right) + C$$

$$\boxed{\int x^5 e^{2x^3} dx = \frac{1}{12} e^{2x^3}(2x^3 - 1) + C}$$

060

$$\int (x^2 - x)e^{-x}dx$$

It is selected $u = (x^2 - x)$ y $dv = e^{-x}\ dx$.

$$u = (x^2 - x) \;\rightarrow\; du = (2x - 1)dx$$
$$dv = e^{-x}\ dx \;\rightarrow\; \int dv = \int e^{-x}\ dx \;\rightarrow\; v = -e^{-x}$$

Therefore, the integration by parts produces

$$\int (x^2 - x)e^{-x}dx = -(x^2 - x)e^{-x} + \int (2x - 1)\,e^{-x}dx \quad \textbf{Equation 1}$$

The integral is solved $\int (2x - 1)\,e^{-x}dx$

It is selected $u = (2x - 1)$ y $dv = e^{-x}\ dx$.

$$u = (2x - 1) \;\rightarrow\; du = 2dx$$
$$dv = e^{-x}\ dx \;\rightarrow\; \int dv = \int e^{-x}\ dx \;\rightarrow\; v = -e^{-x}$$

Therefore, the integration by parts produces

$$\int (2x - 1)\,e^{-x}dx = -(2x - 1)e^{-x} + 2\int e^{-x}dx \quad \textbf{Equation 2}$$

Substitute equation 2 in equation 1

$$\int (x^2 - x)e^{-x}dx = -(x^2 - x)e^{-x} - (2x - 1)e^{-x} + 2\int e^{-x}dx$$

$$\int (x^2 - x)e^{-x}dx = -(x^2 - x)e^{-x} - (2x - 1)e^{-x} - 2e^{-x} + C$$

$$\boxed{\int (x^2 - x)e^{-x}dx = -e^{-x}(x^2 + x + 1) + C}$$

In the following exercises you will find integrals applying the integration technique for trigonometric functions

$$\int sen^3x\, cos^2x\, dx$$

061

Since the power of the sinus is odd and positive, when decomposing the bosom cube a senx is conserved and the other factors are converted to cosines.

$$\int sen^3x\, cos^2x\, dx = \int sen^2x\, cos^2x\, (sen\, x)\, dx$$

$$\int sen^3x\, cos^2x\, dx = \int (1 - cos^2x)cos^2x\, (sen\, x)\, dx)$$

$$\int sen^3x\, cos^2x\, dx = \int (cos^2x - cos^4x)\, (sen\, x)\, dx)$$

$$\int sen^3x\, cos^2x\, dx = \int cos^2x\, sen\, x\, dx - \int cos^4x\, sen\, x\, dx$$

Then the next substitution is made

$$u = cos\, x \;\rightarrow\; du = -sen\, x\, dx$$

$$\int sen^3x\, cos^2x\, dx = -\int u^2 du + \int u^4 dx = -\frac{cos^3x}{3} + \frac{cos^5x}{5} + c$$

$$\boxed{\int sen^3x\, cos^2x\, dx = -\frac{cos^3x}{3} + \frac{cos^5x}{5} + c}$$

Notes

- The *senx* that is in parentheses in the first step is the factor that is conserved.
- To convert the other factors to breasts, the trigonometric identity is used Pythagorean $sen^2x + cos^2x = 1$.

062

$$\int sen^2x \, cos^3x \, dx$$

Since the power of the cosine is odd and positive, by decomposing the cosine cube a cos x is conserved and the other factors are converted to sines.

$$\int sen^2x \, cos^3x \, dx = \int sen^2x \, cos^2x \, (cos \, x) \, dx$$

$$\int sen^2x \, cos^3x \, dx = \int sen^2x(1 - sen^2x) \, (cos \, x) \, dx$$

$$\int sen^2x \, cos^3x \, dx = \int (sen^2x - sen^4x) \, (cos \, x) \, dx$$

$$\int sen^2x \, cos^3x \, dx = \int sen^2x \, cos \, x \, dx - \int sen^4x \, cos \, x \, dx$$

Then the next substitution is made

$$u = sen \, x \rightarrow du = cos \, x \, dx$$

$$\int sen^2x \, cos^3x \, dx = \int u^2 du - \int u^4 dx$$

$$\int sen^2x \, cos^3x \, dx = \frac{sen^3x}{3} - \frac{sen^5x}{5} + c$$

$$\int sen^2x \, cos^3x \, dx = \frac{cos^3x}{3} - \frac{cos^5x}{5} + c$$

Notas

- The cosx that is in parentheses in the first step is the factor that is conserved.
- To convert the other factors to breasts, the trigonometric identity is used Pythagorean $sen^2x + cos^2x = 1$.

$$\int sen^4x\ cos^4x\ dx$$

Now we proceed to rewrite the integral

$$\int sen^4x\ cos^4x\ dx = (sen\ x\ cos\ x)^4\ dx$$

$$sen\ 2x = 2sen\ x\ cos\ x \quad \rightarrow \quad sen\ x\ cos\ x = \frac{sen\ 2x}{2}$$

$$\int (sen\ x\ cos\ x)^4\ dx = \int \left(\frac{sen\ 2x}{2}\right)^4 dx = \frac{1}{16}\int sen^4 2x\ dx$$

$$\int (sen\ x\ cos\ x)^4\ dx = \frac{1}{16}\int (sen^2 2x)^2\ dx = \frac{1}{16}\int \left(\frac{1 - cos\ 4x}{2}\right)^2 dx$$

$$\int (sen\ x\ cos\ x)^4\ dx = \frac{1}{64}\int (1 - cos\ 4x)^2\ dx = \frac{1}{64}\int (1 - 2\ cos\ 4x + cos^2 4x)dx$$

$$\int (sen\ x\ cos\ x)^4\ dx = \frac{1}{64}\int dx - \frac{1}{32}\int cos\ 4x dx + \frac{1}{64}\int cos^2 4x dx$$

$$\int (sen\ x\ cos\ x)^4\ dx = \frac{1}{64}\int dx - \frac{1}{32}\int cos\ 4x dx + \frac{1}{64}\int \left(\frac{1 - cos\ 8x}{2}\right) dx$$

$$\int (sen\ x\ cos\ x)^4\ dx = \frac{1}{64}\int dx - \frac{1}{32}\int cos\ 4x dx + \frac{1}{128}\int dx - \frac{1}{128}\int cos8x\ dx$$

$$\int sen^4x\ cos^4x\ dx = \frac{1}{64}x - \frac{1}{128}sen\ 4x + \frac{1}{128}x - \frac{1}{1024}sen\ 8x + C$$

$$\boxed{\int sen^4x\ cos^4x\ dx = \frac{1}{128}\left(3x - sen\ 4x - \frac{1}{8}sen\ 8x\right) + C}$$

064

$$\int tan^4x \; sec^4x \; dx$$

As the power of the secant is even and positive, the secant is decomposed to the fourth in square secant by square secant, a square secant factor is conserved and the remaining factors are converted into tangents.

$$\int tan^4x \; sec^4x \; dx = \int tan^4x \; sec^2x \; (sec^2x)dx$$

$$\int tan^4x \; sec^4x \; dx = \int tan^4x \; (1 + tan^2x)(sec^2x) \; dx$$

$$\int tan^4x \; sec^4x \; dx = \int (tan^4x + tan^6x)(sec^2x) \; dx$$

$$\int tan^4x \; sec^4x \; dx = \int tan^4x \; sec^2x \; dx + \int tan^6x \; sec^2x \; dx$$

Then the next substitution is made $u = \tan x \;\rightarrow\; du = sec^2x \; dx$

$$\int tan^4x \; sec^4x \; dx = \int u^4du + \int u^6du$$

$$\boxed{\int tan^4x \; sec^4x \; dx = \frac{1}{5}tan^5x + \frac{1}{7}tan^7x + c}$$

065

$$\int tan^2x \; sec^2x \; dx$$

The next substitution is made

$$u = \tan x \;\rightarrow\; du = \; sec^2x \; dx$$

$$\int tan^2x \; sec^2x \; dx = \int u^2du \;\rightarrow\; \boxed{\int tan^2x \; sec^2x \; dx = \frac{tan^3x}{3} + c}$$

$$\int sec^3 x \tan x \, dx$$

066

Since the power of the secant is odd and positive, it is decomposed in square secant by drying, keeping a drying factor for tangent and converting the remaining factors to secant if they exist.

$$\int sec^3 x \tan x \, dx = \int sec^2 x \, (\sec x \tan x) \, dx$$

Then the next substitution is made

$$u = \sec x \rightarrow du = \sec x \tan x \, dx$$

Consequently, you have

$$\int sec^3 x \tan x \, dx = \int u^2 du = \frac{u^3}{3} + C = \frac{sec^3 x}{3} + C$$

$$\boxed{\int \mathbf{sec^3 x \tan x \, dx} = \frac{\mathbf{sec^3 x}}{\mathbf{3}} + \mathbf{c}}$$

$$\int \tan^2 x \, dx$$

067

There are no drying factors. Since the power of the tangent is even and positive, convert the square tangent factor to square secant.

$$\int \tan^2 x \, dx = \int (sec^2 x - 1) \, dx = \int sec^2 x dx - \int dx$$

$$\boxed{\int \tan^2 x \, dx = \tan x - x + c}$$

068

$$\int \tan^3 2t \; sec^3 2t \; dt$$

Since the power of the secant is odd and positive, it is decomposed in a secant by a square secant, preserving a secant factor per tangent and converting the remaining factors to secant.

$$\int \tan^3 2t \; sec^3 2t \; dt = \int \tan^2 2t \; sec^2 2t(sec2t \; tan2t)dt$$

$$\int \tan^3 2t \; sec^3 2t \; dt = \int (sec^2 2t - 1)sec^2 2t(sec2t \; tan2t)dt$$

$$\int \tan^3 2t \; sec^3 2t \; dt = \int sec^4 2t(sec2t \; tan2t)dt - \int sec^2 2t(sec2t \; tan2t)dt$$

$$u = \sec 2t \rightarrow du = 2\sec 2t \tan 2t \; dt$$

$$\int \tan^3 2t \; sec^3 2t \; dt = \frac{1}{2}\int u^4 du - \frac{1}{2}\int u^2 \; du$$

$$\boxed{\int \tan^3 2t \; sec^3 2t \; dt = \frac{1}{10}sec^5 2t - \frac{1}{6}sec^3 + c}$$

069

$$\int \tan^{\frac{3}{2}}x \; sec^4 x \; dx$$

$$\int \tan^{\frac{3}{2}}x \; sec^4 x \; dx = \int \tan^{\frac{3}{2}}x \; sec^2 x(sec^2 x) \; dx = \int \tan^{\frac{3}{2}}x \; (1 + tan^2 x)(sec^2 x) \; dx$$

$$\int \tan^{\frac{3}{2}}x \; sec^4 x \; dx = \int \tan^{\frac{3}{2}}x \; sec^2 x dx + \int \tan^{\frac{7}{2}}x \; sec^2 x dx$$

Se utiliza la siguiente sustitución: $u = \tan x \rightarrow du = sec^2 x dx$

$$\int \tan^{\frac{3}{2}}x \; sec^4 x \; dx = \int u^{\frac{3}{2}}du + \int u^{\frac{7}{2}}du \qquad \boxed{\int \tan^{\frac{3}{2}}x \; sec^4 x \; dx = \frac{2tan^{\frac{5}{2}}x}{5} + \frac{2tan^{\frac{9}{2}}x}{9} + c}$$

$$\int \frac{tan^2x}{sec^5x}dx$$

In the integral we have a quotient, which indicates that the integral must be rewritten in the following terms: The tangent is expressed as a function of sine and cosine and $\frac{1}{secante}$ depending on cosine.

$$\int \frac{tan^2x}{sec^5x}dx = \int \frac{sen^2x}{cos^2x}\,cos^5dx$$

$$\int \frac{tan^2x}{sec^5x}dx = \int sen^2x\,cos^3x\,dx$$

$$\int \frac{tan^2x}{sec^5x}dx = \int sen^2x\,cos^2x\,(\cos x)\,dx$$

$$\int \frac{tan^2x}{sec^5x}dx = \int sen^2x\,(1-sen^2x)\,(\cos x)\,dx$$

$$\int \frac{tan^2x}{sec^5x}dx = \int sen^2x\,cosx\,dx - \int sen^4x\,cosx\,dx$$

Then the next substitution is made

$$u = sen\,x \;\rightarrow\; du = cosx\,dx$$
$$\int \frac{tan^2x}{sec^5x}dx = \int u^2\,du - \int u^4 du$$

$$\boxed{\int \frac{tan^2x}{sec^5x}dx = \frac{sen^3x}{3} - \frac{sen^5x}{5} + c}$$

071

$$\int \frac{tan^2x}{\sec x}dx$$

In the integral we have a quotient, which indicates that the integral must be rewritten in the following terms: The tangent is expressed as a function of sine and cosine and $\frac{1}{\sec x}$ depending on cosine.

$$\int \frac{tan^2x}{\sec x}dx = \int \frac{sen^2x}{cos^2x}\cos x\, dx$$

$$\int \frac{tan^2x}{\sec x}dx = \int \frac{sen^2x}{\cos x}dx$$

$$\int \frac{tan^2x}{\sec x}dx = \int \frac{(1-cos^2x)}{\cos x}dx$$

$$\int \frac{tan^2x}{\sec x}dx = \int \frac{1}{\cos x}dx - \int \frac{cos^2x}{\cos x}dx$$

$$\int \frac{tan^2x}{\sec x}dx = \int \sec x\, dx - \int \cos x\, dx$$

$$\boxed{\int \frac{tan^2x}{\sec x}dx = \ln|\sec x + \tan x| - sen\, x + c}$$

Note: Another procedure to solve the integral is to apply the trigonometric identity $1 + tan^2x = sec^2x$. Then it clears tan^2x and it is replaced in the integral. It is suggested to the reader to solve the integral and to corroborate the result with the previous one.

$$\int tan^3 \, 3x \, dx$$

There are no drying factors. The power of the tangent is not even. What is appropriate is to decompose the $tan^3 3x$ in $tan^2 3x \tan 3x$ and work with the trigonometric identity $1 + tan^2 x = sec^2 x$.

$$\int tan^3 \, 3x \, dx = \int tan^2 \, 3x \tan 3x \, dx$$

$$\int tan^3 \, 3x \, dx = \int (sec^2 3x - 1) \tan 3x \, dx$$

$$\int tan^3 \, 3x \, dx = \int \tan 3x \, sec^2 3x \, dx - \int \tan 3x \, dx$$

$$\int tan^3 \, 3x \, dx = \int \tan 3x \, sec^2 3x \, dx - \int \frac{sen \, 3x}{cos \, 3x} \, dx$$

The following substitutions are made

$$u = \tan 3x \;\to\; du = 3sec^2 3x \, dx$$

$$w = \cos 3x \;\to\; dw = -3sen \, 3x \, dx$$

Consequently, you have

$$\int tan^3 \, 3x \, dx = \frac{1}{3} \int u \, du + \frac{1}{3} \int \frac{dw}{w}$$

$$\boxed{\int tan^3 \, 3x \, dx = \frac{tan^2 3x}{6} + \frac{1}{3} ln|cos \, 3x| + c}$$

Note: In step three the second integral is immediate. However, a substitution was used in the problem.

073

$$\int tan^3 \frac{\pi x}{2} \; sec^2 \frac{\pi x}{2} \; dx$$

The theory says that if the power of the secant is even and positive, keep a square secant factor and convert the remaining factors into tangents. As you can see, the integral presents a square secant factor, and there are no remaining factors to turn them into tangents. What is done then? What is appropriate is to convert the square secant into a square tangent.

$$\int tan^3 \frac{\pi x}{2} \; sec^2 \frac{\pi x}{2} \; dx = \int tan^3 \frac{\pi x}{2} \left(1 + tan^2 \frac{\pi x}{2} \right) dx$$

$$\int tan^3 \frac{\pi x}{2} \; sec^2 \frac{\pi x}{2} \; dx = \int tan^3 \frac{\pi x}{2} dx + \int tan^5 \frac{\pi x}{2} dx \quad \textbf{\textit{Equation}} \; 1$$

Then we proceed to solve each integral, and adding the results we get the result of the integral problem.

$$\int tan^3 \frac{\pi x}{2} dx = tan^2 \frac{\pi x}{2} \; tan \frac{\pi x}{2} dx = \int \left(sec^2 \frac{\pi x}{2} - 1 \right) tan \frac{\pi x}{2} dx$$

$$\int tan^3 \frac{\pi x}{2} dx = \int tan \frac{\pi x}{2} \; sec^2 \frac{\pi x}{2} \; dx - \int tan \frac{\pi x}{2} dx$$

The following substitution is made

$$u = tan \frac{\pi x}{2} \rightarrow du = \frac{\pi}{2} sec^2 \frac{\pi x}{2} \; dx$$

$$\int tan^3 \frac{\pi x}{2} dx = \frac{2}{\pi} \int u \, du - \int tan \frac{\pi x}{2} dx$$

$$\int tan^3 \frac{\pi x}{2} dx = \frac{2}{\pi} \frac{tan^2 \frac{\pi x}{2}}{2} - \frac{2}{\pi} ln \left| sec \frac{\pi x}{2} \right| + C$$

$$\int tan^3 \frac{\pi x}{2} dx = \frac{1}{\pi} tan^2 \frac{\pi x}{2} - \frac{2}{\pi} ln \left| sec \frac{\pi x}{2} \right| + C \quad \textbf{\textit{Equation}} \; 2$$

$$\int tan^5 \frac{\pi x}{2} dx = \int tan^2 \frac{\pi x}{2} \, tan^3 \frac{\pi x}{2} \, dx$$

$$\int tan^5 \frac{\pi x}{2} dx = \int \left(sec^2 \frac{\pi x}{2} - 1 \right) tan^3 \frac{\pi x}{2} dx$$

$$\int tan^5 \frac{\pi x}{2} dx = \int tan^3 \frac{\pi x}{2} \, sec^2 \frac{\pi x}{2} \, dx - \int tan^3 \frac{\pi x}{2} dx$$

$$\int tan^5 \frac{\pi x}{2} dx = \int tan^3 \frac{\pi x}{2} \, sec^2 \frac{\pi x}{2} \, dx - \int \left(sec^2 \frac{\pi x}{2} - 1 \right) tan \frac{\pi x}{2} dx$$

$$\int tan^5 \frac{\pi x}{2} dx = \int tan^3 \frac{\pi x}{2} \, sec^2 \frac{\pi x}{2} \, dx - \int tan \frac{\pi x}{2} \, sec^2 \frac{\pi x}{2} \, dx + \int tan \frac{\pi x}{2} dx$$

The following substitution is made

$$u = tan \frac{\pi x}{2} \rightarrow du = \frac{\pi}{2} sec^2 \frac{\pi x}{2} \, dx$$

$$\int tan^5 \frac{\pi x}{2} dx = \frac{2}{\pi} \int u^3 du - \frac{2}{\pi} \int u du + \int tan \frac{\pi x}{2} dx$$

$$\int tan^5 \frac{\pi x}{2} dx = \frac{2}{\pi} \frac{tan^4 \frac{\pi x}{2}}{4} - \frac{2}{\pi} \frac{tan^2 \frac{\pi x}{2}}{2} + \frac{2}{\pi} ln \left| sec \frac{\pi x}{2} \right| + C$$

$$\boxed{\int tan^5 \frac{\pi x}{2} dx = \frac{1}{2\pi} tan^4 \frac{\pi x}{2} - \frac{1}{\pi} tan^2 \frac{\pi x}{2} + \frac{2}{\pi} ln \left| sec \frac{\pi x}{2} \right| + C \; \textbf{Equation } 3}$$

Substituting the result of equation 2 and of equation 3 in equation 1 and canceling the corresponding terms, we obtain the result of the proposed integral.

$$\boxed{\int tan^3 \frac{\pi x}{2} \, sec^2 \frac{\pi x}{2} \, dx = \frac{1}{2\pi} tan^4 \frac{\pi x}{2} + C}$$

074

$$\int \frac{1}{\sec x \tan x}\, dx$$

The integral is rewritten as follows

$$\int \frac{1}{\sec x \tan x}\, dx = \int \frac{1}{\dfrac{1}{\cos x}\dfrac{sen\, x}{\cos x}}\, dx$$

$$\int \frac{1}{\sec x \tan x}\, dx = \int \frac{1}{\dfrac{sen\, x}{\cos^2 x}}\, dx$$

$$\int \frac{1}{\sec x \tan x}\, dx = \int \frac{\cos^2 x}{sen\, x}\, dx$$

$$\int \frac{1}{\sec x \tan x}\, dx = \int \frac{(1 - sen^2 x)}{sen\, x}\, dx$$

$$\int \frac{1}{\sec x \tan x}\, dx = \int \frac{1}{sen\, x}\, dx - \int \frac{sen^2 x}{sen\, x}\, dx$$

$$\int \frac{1}{\sec x \tan x}\, dx = \int csc x\, dx - \int sen x\, dx$$

Both the first integral and the second are immediate

$$\int \frac{1}{\sec x \tan x}\, dx = ln|csc\, x - \cot x| + \cos x + C$$

$$\boxed{\int \frac{1}{\sec x \tan x}\, dx = ln|csc\, x - \cot x| + \cos x + C}$$

$$\int (\tan^4 t - \sec^4 t)\, dt$$

075

The integral is rewritten as follows

$$\int (\tan^4 t - \sec^4 t)\, dt = \int [(\tan^2 t)^2 - (\tan^2 t)^2]\, dt \rightarrow \ Difference\ of\ squares$$

$$\int (\tan^4 t - \sec^4 t)\, dt = \int (\tan^2 t + \sec^2 t)\,(\tan^2 t - \sec^2 t)\, dt$$

Of the trigonometric identity $1 + \tan^2 t = \sec^2 t$ you have

$$\tan^2 t - \sec^2 t = -1$$

As a result, you get

$$\int (\tan^4 t - \sec^4 t)\, dt = -\int (\tan^2 t + \sec^2 t)\, dt$$

Now the square tangent is converted to square secant

$$\int (\tan^4 t - \sec^4 t)\, dt = -\int (\sec^2 t - 1 + \sec^2 t)\, dt$$

$$\int (\tan^4 t - \sec^4 t)\, dt = -\int (2\sec^2 t - 1)\, dt$$

$$\int (\tan^4 t - \sec^4 t)\, dt = -2\int \sec^2 t\, dt + \int dt$$

$$\int (\tan^4 t - \sec^4 t)\, dt = -2\tan t + t + C$$

$$\boxed{\int (\tan^4 t - \sec^4 t)\, dt = -2\tan t + t + C}$$

076

$$\int \frac{cot^3 t}{csc\, t}\, dt$$

The integral is rewritten by expressing the cotangent as a function of the cosine and sine, and the cosecant as a function of the sine.

$$\int \frac{cot^3 t}{csc\, t}\, dt = \int \frac{\frac{cos^3 t}{sen^3 t}}{\frac{1}{sen\, t}} = \int \frac{sen\, t \, cos^3 t}{sen^3 t}\, dt$$

$$\int \frac{cot^3 t}{csc\, t}\, dt = \int \frac{cos^3 t}{sen^2 t}\, dt = \int \frac{cos^2 t \cos t}{sen^2 t}\, dt$$

$$\int \frac{cot^3 t}{csc\, t}\, dt = \int \frac{(1 - sen^2 t)\cos t}{sen^2 t}\, dt$$

$$\int \frac{cot^3 t}{csc\, t}\, dt = \int \frac{\cos t - sen^2 t \cos t}{sen^2 t}\, dt$$

$$\int \frac{cot^3 t}{csc\, t}\, dt = \int \frac{\cos t}{sen^2 t}\, dt - \int \frac{sen^2 t \cos t}{sen^2 t}\, dt$$

$$\int \frac{cot^3 t}{csc\, t}\, dt = \int \frac{\cos t}{sen^2 t}\, dt - \int \cos t \, dt$$

$$\int \frac{cot^3 t}{csc\, t}\, dt = \int \frac{du}{u^2} - \int \cos t \, dt$$

$$\int \frac{cot^3 t}{csc\, t}\, dt = -\frac{1}{sent} - sent + C$$

$$\boxed{\int \frac{cot^3 t}{csc\, t}\, dt = -csct - sent + C}$$

$$\int \frac{sen^2x - cos^2x}{cos\, x}\, dx$$

077

To solve the integral, the following trigonometric identity is required

$cos^2x - sen^2x = 2cos^2x - 1$. If you multiply by less (-) the identity becomes in $sen^2x - cos^2x = 1 - 2cos^2x$.

Substituting in the integral you get

$$\int \frac{sen^2x - cos^2x}{cos\, x}\, dx = \int \frac{1 - 2cos^2x}{cos\, x}\, dx$$

$$\int \frac{sen^2x - cos^2x}{cos\, x}\, dx = \int \frac{1}{cos\, x}\, dx - 2\int \frac{cos^2x}{cos\, x}\, dx$$

As $\dfrac{1}{cos\, x} = sec\, x$ and canceling similar terms you have

$$\int \frac{sen^2x - cos^2x}{cos\, x}\, dx = \int sec\, x\, dx - 2\int cos\, x\, dx$$

As the two integrals are immediate you get

$$\int \frac{sen^2x - cos^2x}{cos\, x}\, dx = ln|sec\, x + \tan x| - 2sen\, x + C$$

Then the result is

$$\boxed{\int \frac{sen^2x - cos^2x}{cos\, x}\, dx = ln|sec\, x + \tan x| - 2sen\, x + C}$$

078

$$\int \frac{1-\sec t}{\cos t - 1}\, dt$$

The integral is rewritten by expressing the secant as a function of the cosine.

$$\int \frac{1-\sec t}{\cos t - 1}\, dt = \int \frac{1-\frac{1}{\cos t}}{\cos t - 1}\, dt = \int \frac{\left(1-\frac{1}{\cos t}\right)\cos t}{(\cos t - 1)\cos t}\, dt$$

It multiplied and divided both the numerator and the denominator by cosine.

$$\int \frac{1-\sec t}{\cos t - 1}\, dt = \int \frac{(\cos t - 1)}{(\cos t - 1)\cos t}\, dt$$

$$\int \frac{1-\sec t}{\cos t - 1}\, dt = \int \frac{1}{\cos t}\, dt = \int \sec t\, dt$$

$$\boxed{\int \frac{1-\sec t}{\cos t - 1}\, dt = \ln|\sec t + \tan t| + C}$$

079

$$\int \sec 3x\, dx$$

The integrand is multiplied and divided by $\sec x + \tan 3x$

$$\int \sec 3x\, dx = \int \frac{\sec 3x(\sec 3x + \tan 3x)}{(\sec 3x + \tan 3x)}\, dx = \int \frac{\sec^2 3x + \sec 3x \tan 3x}{\sec 3x + \tan 3x}\, dx$$

The following substitution is made

$$u = \sec 3x + \tan 3x \to du = (3\sec 3x \tan 3x + 3\sec^2 3x)\,dx$$

$$\int \sec 3x = \frac{1}{3}\int \frac{du}{u} \to \int \sec 3x = \frac{1}{3}\ln|u| + C$$

$$\boxed{\int \sec 3x = \frac{1}{3}\ln|\sec 3x + \tan 3x| + C}$$

$$\int \tan^3 3x \sec 3x \, dx$$

The strategies studied in the theoretical foundation do not apply to this integral. What is appropriate is to decompose the tangent to obtain the secx factor tanx.

$$\int \tan^3 3x \sec 3x \, dx = \int \tan^2 3x \sec 3x \tan 3x \, dx$$

$$\int \tan^3 3x \sec 3x \, dx = \int (sec^2 3x - 1) \sec 3x \tan 3x \, dx$$

$$\int \tan^3 3x \sec 3x \, dx = \int (sec^2 3x) \sec 3x \tan 3x \, dx - \int \sec 3x \tan 3x dx$$

The following substitution is made

$$u = \sec 3x \;\rightarrow\; du = 3 \sec 3x \tan 3x \, dx$$

And multiply and divide the second integral by three.

$$\int \tan^3 3x \sec 3x \, dx = \frac{1}{3}\int u^2 du - \frac{1}{3}\int 3 \sec 3x \tan 3x \, dx$$

$$\int \tan^3 3x \sec 3x \, dx = \frac{1}{3}\int u^2 du - \frac{1}{3}\int 3 \sec 3x \tan 3x \, dx$$

$$\int \tan^3 3x \sec 3x \, dx = \frac{1}{3}\int u^2 du - \frac{1}{3}\int du$$

$$\int \tan^3 3x \sec 3x \, dx = \frac{1}{9}u^3 - \frac{1}{3}u + C$$

$$\int \tan^3 3x \sec 3x \, dx = \frac{1}{9}sec^3 3x - \frac{1}{3}\sec 3x + C$$

$$\boxed{\int \tan^3 3x \sec 3x \, dx = \frac{1}{9}sec^3 3x - \frac{1}{3}\sec 3x + C}$$

081

$$\int \cot^3 x \ \csc^4 x \ dx$$

As the strength of the cosecant is even and positive, a square cosecant is retained and the remaining factors are converted to cotangents.

$$\int \cot^3 x \ \csc^4 x \ dx = \int \cot^3 x \ \csc^2 x \ \csc^2 x \ dx = \int \cot^3 x (1 + \cot^2 x) \ \csc^2 x \ dx$$

$$\int \cot^3 x \ \csc^4 x \ dx = \int \cot^3 x \ \csc^2 x \ dx + \int \cot^5 x \ \csc^2 x \ dx$$

Then the next substitution is made

$$u = \cot x \ \rightarrow \ du = -\csc^2 x \ dx$$

$$\int \cot^3 x \ \csc^4 x \ dx = -\int u^3 du - \int u^5 du = -\frac{1}{4}\cot^4 x - \frac{1}{6}\cot^6 x + C$$

$$\boxed{\int \cot^3 x \ \csc^4 x \ dx = -\frac{1}{4}\cot^4 x - \frac{1}{6}\cot^6 x + C}$$

082

$$\int \frac{1 - \cos 2x}{1 + \cos 2x} \ dx$$

Multiplies and divides by 1/2 the integral to obtain the identity trigonometric of the mean angle of the sine and the cosine.

$$\int \frac{1 - \cos 2x}{1 + \cos 2x} \ dx = \int \frac{\dfrac{1 - \cos 2x}{2}}{\dfrac{1 + \cos 2x}{2}} \ dx = \int \frac{sen^2 x}{\cos^2 x} \ dx = \int \tan^2 x \ dx$$

$$\int \frac{1 - \cos 2x}{1 + \cos 2x} \ dx = \int (\sec^2 x - 1) \ dx = \int \sec^2 x \ dx - \int dx$$

$$\boxed{\int \frac{1 - \cos 2x}{1 + \cos 2x} \ dx = \tan x - x + c}$$

$$\int sec^3\pi x \; dx$$

083

If the integral is of the form $\int sec^m x \; dx$ where m is odd and positive, use integration by parts. The integral is rewritten by decomposing the blotting cube in square blotter by blotter.

$$\int sec^3\pi x \; dx = \int sec^2\pi x \; sec \; \pi x \; dx$$

$$u = \; sec \, \pi x \rightarrow du = \pi \, sec \, \pi x \, \tan \pi x \; dx$$

$$dv = \; sec^2\pi x \; dx \; \rightarrow \int dv = \int sec^2\pi x \; dx \; \rightarrow \; v = \; \frac{1}{\pi} \tan \pi x$$

$$\int sec^3\pi x \; dx = \frac{1}{\pi} sec \, \pi x \tan \pi x - \int sec \, \pi x \tan \pi x \tan \pi x \; dx$$

$$\int sec^3\pi x \; dx = \frac{1}{\pi} sec \, \pi x \tan \pi x - \int sec \, \pi x \; tan^2\pi x \; dx$$

$$\int sec^3\pi x \; dx = \frac{1}{\pi} sec \, \pi x \tan \pi x - \int sec \, \pi x \, (sec^2\pi x - 1) \; dx$$

$$\int sec^3\pi x \; dx = \frac{1}{\pi} sec \, \pi x \tan \pi x - \int sec^3\pi x \; dx + \int sec \, \pi x \; dx$$

$$\int sec^3\pi x \; dx + \int sec^3\pi x \; dx = \frac{1}{\pi} sec \, \pi x \tan \pi x + \int sec \, \pi x \; dx$$

$$2 \int sec^3\pi x \; dx = \frac{1}{\pi} sec \, \pi x \tan \pi x + \frac{1}{\pi} ln|sec \, \pi x + \tan \pi x| + C$$

$$\boxed{\int sec^3\pi x \; dx = \frac{1}{2\pi} \left[sec \, \pi x \tan \pi x + ln|sec \, \pi x + \tan \pi x| \right] + C}$$

084

$$\int sen^4x \; cos^2x \; dx$$

The strategies proposed in the theoretical foundation of the book do not fit this integral. The coming is to decompose the sine to the fourth in square sine per square sine.

$$\int sen^4x \; cos^2x \; dx = \int (sen^2x \; cos^2x)sen^2dx \; = \int (sen \; x \cos x)^2 \; sen^2x \; dx$$

Of the trigonometric identity $sen \; 2x = 2sen \; x \cos x \; it \; clears \; sen \; x \cos x$ and considering
$$sen^2x = \frac{1 - \cos 2x}{2}$$

$$\int sen^4x \; cos^2x \; dx = \int \left(\frac{sen \; 2x}{2}\right)^2 \left(\frac{1 - \cos 2x}{2}\right) dx$$

$$\int sen^4x \; cos^2x \; dx = \frac{1}{8}\int sen^2 2x \; (1 - \cos 2x)dx$$

$$\int sen^4x \; cos^2x \; dx = \frac{1}{8}\int sen^2 2x \; dx - \frac{1}{8}\int sen^2 2x \cos 2x \; dx$$

$$\int sen^4x \; cos^2x \; dx = \frac{1}{8}\int \frac{1 - \cos 4x}{2} \; dx - \frac{1}{8}\int sen^2 2x \cos 2x \; dx$$

$$\int sen^4x \; cos^2x \; dx = \frac{1}{16}\int dx - \frac{1}{16}\int \cos 4x \; dx - \frac{1}{8}\int sen^2 2x \cos 2x \; dx$$

$$\int sen^4x \; cos^2x \; dx = \frac{1}{16} x - \frac{1}{64} sen \; 4x - \frac{1}{16}\int u^2 du$$

$$\int sen^4x \; cos^2x \; dx = \frac{1}{16} x - \frac{1}{64} sen \; 4x - \frac{1}{48} sen^3 2x + C$$

$$\boxed{\int sen^4x \; cos^2x \; dx = \frac{1}{16} x - \frac{1}{64} sen \; 4x - \frac{1}{48} sen^3 2x + C}$$

$$\int \sec^6 3x\, dx$$

085

The integral is rewritten by decomposing the secant at six in blotting to the fourth and blotting in squares.

$$\int \sec^6 3x\, dx = \int \sec^4 3x \sec^2 3x\, dx = \int (1 + tan^2 3x)^2\ \sec^2 3x\, dx$$

$$\int \sec^6 3x\, dx = \int (1 + 2tan^2 3x + tan^4 3x)\sec^2 3x\, dx$$

$$\int \sec^6 3x\, dx = \int \sec^2 3x\, dx + 2 \int tan^2 3x \sec^2 3x\, dx + \int tan^4 3x \sec^2 3x\, dx$$

$$\int \sec^6 3x\, dx = \frac{1}{3}\tan 3x + \frac{2}{3}\int u^2\, du + \frac{1}{3}\int u^4 du$$

$$\boxed{\int \sec^6 3x\, dx = \frac{1}{3}\left[\tan 3x + \frac{2}{3}\tan^3 3x + \frac{1}{5}\tan^5 3x\right] + C}$$

$$\int sen\, 3x \cos 2x\, dx$$

086

To solve the integral we use the identity of the product sine-cosine sum and initially solving the integrand, we have

$$sen\, \alpha \cos \beta = \frac{1}{2}[sen\,(\alpha - \beta) + sen\,(\alpha + \beta)]$$

$$sen\, 3x \cos 2x = \frac{1}{2}[sen\,(3x - 2x) + sen\,(3x + 2x)] = \frac{1}{2}[sen x + sen 5x]$$

$$\int sen\, 3x \cos 2x\, dx = \frac{1}{2}\int sen\, x\, dx + \frac{1}{2}\int sen\, 5x\, dx = -\frac{1}{2}\cos x - \frac{1}{10}\cos 5x + C$$

$$\boxed{\int sen\, 3x \cos 2x\, dx = -\frac{1}{10}(5\cos x + \cos 5x) + C}$$

117

087

$$\int sen\ \theta\ sen\ 3\theta\ d\theta$$

To solve the integral the following identity is used

$$sen\ \alpha\ sen\ \beta = \frac{1}{2}[\cos(\alpha - \beta) - \cos(\alpha + \beta)]$$

The integrando is solved initially

$$sen\ \theta\ sen\ 3\theta = \frac{1}{2}[\cos(\theta - 3\theta) - \cos(\theta + 3\theta)] = \frac{1}{2}[\cos(-2\theta) - \cos(4\theta)]$$

To solve the $\cos(-2\theta)$ the reduction formula is used

$$\cos(-x) = \cos x \rightarrow \cos(-2\theta) = \cos 2\theta$$

$$sen\ \theta\ sen\ 3\theta = \frac{1}{2}[\cos(2\theta) - \cos(4\theta)]$$

$$\int sen\ \theta\ sen\ 3\theta\ d\theta = \frac{1}{2}\int(\cos 2\theta - \cos 4\theta)d\theta$$

$$\int sen\ \theta\ sen\ 3\theta\ d\theta = \frac{1}{2}\int \cos 2\theta\ d\theta - \frac{1}{2}\int \cos 4\theta\ d\theta$$

$$\int sen\ \theta\ sen\ 3\theta\ d\theta = \frac{1}{4}sen\ 2\theta - \frac{1}{8}sen\ 4\theta + C$$

$$\int sen\ \theta\ sen\ 3\theta\ d\theta = \frac{1}{8}(2sen\ 2\theta - sen\ 4\theta) + C$$

$$\boxed{\int \boldsymbol{sen\ \theta\ sen\ 3\theta\ d\theta} = \frac{1}{8}(\boldsymbol{2sen\ 2\theta - sen\ 4\theta}) + \boldsymbol{C}}$$

$$\int \cos 4\theta \cos(-3\theta)d\theta$$

088

To solve the integral the following identity is used

$$cos\ \alpha\ cos\ \beta = \frac{1}{2}[\cos(\alpha - \beta) + \cos(\alpha + \beta)]$$

The integral is rewritten taking into consideration that

$$\cos(-3\theta) = cos\ 3\theta \quad \text{therefore, you have}$$

$$\int \text{Cos } 4\theta \cos(-3\theta)d\theta = \int \text{Cos } 4\theta \cos 3\theta\ d\theta$$

The integrando is solved initially

$$\cos 4\theta \cos 3\theta = \frac{1}{2}[\cos(4\theta - 3\theta) + \cos(4\theta + 3\theta)] = \frac{1}{2}[\cos(\theta) + \cos(7\theta)]$$

$$\int \cos 4\theta \cos(-3\theta)d\theta = \int \frac{1}{2}(\cos\theta + \cos 7\theta)\ d\theta$$

$$\int \cos 4\theta \cos(-3\theta)d\theta = \frac{1}{2}\int (\cos\theta + \cos 7\theta)\ d\theta$$

$$\int \cos 4\theta \cos(-3\theta)d\theta = \frac{1}{2}\int \cos\theta\ d\theta + \frac{1}{2}\int \cos 7\theta\ d\theta$$

$$\int \cos 4\theta \cos(-3\theta)d\theta = \frac{1}{2}sen\ \theta + \frac{1}{14}sen\ 7\theta + C$$

$$\int \cos 4\theta \cos(-3\theta)d\theta = \frac{1}{14}(7sen\ \theta + sen\ 7\theta) + C$$

$$\boxed{\int \mathbf{\cos 4\theta \cos(-3\theta)d\theta} = \frac{1}{14}(\mathbf{7sen\ \theta + sen\ 7\theta}) + C}$$

089

$$\int sen(-4x)\cos 3x\,dx$$

The integral is rewritten taking into consideration that $sen(-4x) = -sen\,4x$.

$$\int sen(-4x)\cos 3x\,dx = \int -sen(4x)\cos 3x\,dx = -\int sen(4x)\cos 3x\,dx$$

Now we proceed to solve the integrando using the following identity

$$sen\,4x\cos 3x = \frac{1}{2}[sen\,(4x-3x)+sen\,(4x+3x)] = \frac{1}{2}[sen\,x + sen\,7x]$$

$$\int sen(-4x)\cos 3x\,dx = -\int \frac{1}{2}[sen\,x + sen\,7x]dx$$

$$\int sen(-4x)\cos 3x\,dx = -\frac{1}{2}\int [sen\,x + sen\,7x]dx$$

$$\int sen(-4x)\cos 3x\,dx = -\frac{1}{2}\int sen\,x\,dx - \frac{1}{2}\int sen\,7x\,dx$$

$$\int sen(-4x)\cos 3x\,dx = \frac{1}{2}\cos x + \frac{1}{14}\cos 7x + C$$

$$\int sen(-4x)\cos 3x\,dx = \frac{1}{14}\,(7\cos x + \cos 7x) + C$$

$$\boxed{\int \mathbf{sen(-4x)\cos 3x\,dx} = \frac{1}{14}\,(\mathbf{7\cos x + \cos 7x}) + C}$$

$$\int \frac{dx}{sen^2x\,cos^4x}$$

The integral is rewritten

$$\int \frac{dx}{sen^2x\,cos^4x} = \int \frac{1}{sen^2x\,cos^4x}\,dx$$

As $sen^2x + cos^2 = 1$ you have

$$\int \frac{dx}{sen^2x\,cos^4x} = \int \frac{sen^2x + cos^2x}{sen^2x\,cos^4x}\,dx$$

$$\int \frac{dx}{sen^2x\,cos^4x} = \int \frac{sen^2x}{sen^2x\,cos^4x}\,dx + \int \frac{cos^2x}{sen^2x\,cos^4x}\,dx$$

$$\int \frac{dx}{sen^2x\,cos^4x} = \int \frac{dx}{cos^4x} + \int \frac{dx}{sen^2x\,cos^2x} = \int sec^4x\,dx + \int \frac{dx}{(sen\,x\,cos\,x)^2}$$

$$\int \frac{dx}{sen^2x\,cos^4x} = \int sec^4x\,dx + \int \frac{dx}{\left(\frac{sen\,2x}{2}\right)^2} = \int sec^4x\,dx + 4\int \frac{dx}{sen^2 2x}$$

$$\int \frac{dx}{sen^2x\,cos^4x} = \int sec^4x\,dx + 4\int csc^2 2x\,dx = \int sec^2x\,sec^2x\,dx + 4\int csc^2 2x\,dx$$

$$\int \frac{dx}{sen^2x\,cos^4x} = \int (1 + tan^2x)\,sec^2x\,dx + 4\int csc^2 2x\,dx$$

$$\int \frac{dx}{sen^2x\,cos^4x} = \int sec^2x\,dx + \int tan^2x\,sec^2x\,dx + 4\int csc^2 2x\,dx$$

$$\boxed{\int \frac{dx}{sen^2x\,cos^4x} = \tan x + \frac{tan^3x}{3} - 2\cot 2x + C}$$

**In the following exercises
are the integrals applying the technique
of trigonometric substitution.**

$$\int \frac{1}{\sqrt{16 - x^2}}\, dx$$

091

The integral is rewritten to determine that the form is $\sqrt{a^2 - u^2}$.

$$\int \frac{1}{\sqrt{16 - x^2}}\, dx = \int \frac{1}{\sqrt{4^2 - x^2}}\, dx$$

Where: $a = 4$, $x = 4 sen\,\theta$, $dx = 4\cos\theta d\theta$, $\sqrt{4^2 - x^2} = 4\cos\theta$

Now we proceed to make substitutions and integrate

$$\int \frac{1}{\sqrt{16 - x^2}}\, dx = \int \frac{1}{4cos\theta}\, 4cos\theta d\theta = \int d\theta = \theta + C$$

To return to the original variable, we proceed to construct the right triangle

As $x = 4 sen\theta$ *you have* $sen\theta = \frac{x}{4}$

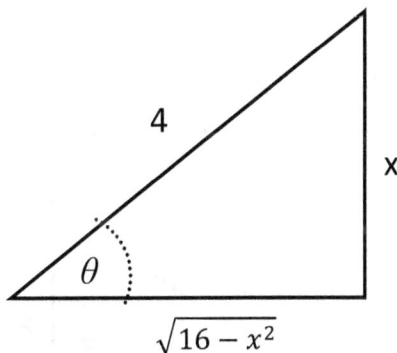

$$4$$

$$x$$

$$\theta$$

$$\sqrt{16 - x^2}$$

Clearing θ: $sen\theta = \frac{x}{4}\ \rightarrow\ \theta = arcsen\,\frac{x}{4}$

$$\boxed{\int \frac{1}{\sqrt{16 - x^2}}\, dx = arcsen\,\frac{x}{4} + C}$$

092

$$\int \sqrt{16 - 4x^2} \ dx$$

The integral is rewritten to determine that the form is: $\sqrt{a^2 - u^2}$

$$\int \sqrt{16 - 4x^2} \ dx = \int \sqrt{4^2 - (2x)^2} \ dx$$

Where: $a = 4$, $2x = 4sen \ \theta$, $x = 2sen \ \theta$, $dx = 2\cos \theta d\theta$, $\sqrt{4^2 - (2x)^2} = 4\cos\theta$

Now we proceed to make substitutions and integrate

$$\int \sqrt{16 - 4x^2} \ dx = \int 4\cos\theta \ 2\cos\theta d\theta = 8\int \cos^2 \ d\theta$$

$$\int \sqrt{16 - 4x^2} \ dx = 8\int \frac{1 + \cos2\theta}{2} d\theta = 4\int (1 + \cos2\theta) \ d\theta$$

$$\int \sqrt{16 - 4x^2} \ dx = 4\int d\theta + 4\int \cos2\theta \ d\theta = 4\theta + 2sen2\theta + C$$

To return to the original variable, we proceed to construct the right triangle

As $2x = 4sen\theta$ you have: $sen\theta = \frac{x}{2}$ $\rightarrow$ $\theta = arcsen\frac{x}{2}$

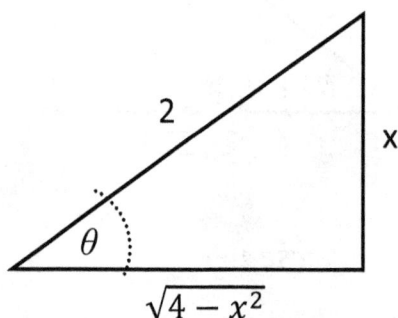

$$\int \sqrt{16 - 4x^2} \ dx = 4\theta + 4sen\theta \ \cos\theta + C = 4arcsen\frac{x}{2} + 4\frac{x}{2}\frac{\sqrt{4 - x^2}}{2} + C$$

$$\boxed{\int \sqrt{16 - 4x^2} \ dx = 4arcsen\frac{x}{2} + x\sqrt{4 - x^2} + C}$$

$$\int \frac{1}{\sqrt{x^2 - 9}}\, dx$$

093

The integral is rewritten to determine that the form is: $\sqrt{u^2 - a^2}$

$$\int \frac{1}{\sqrt{x^2 - 9}}\, dx = \int \frac{1}{\sqrt{x^2 - 3^2}}\, dx$$

Where: $a = 3$, $x = 3\sec\theta$, $\sqrt{x^2 - 9} = 3\tan\theta$, $dx = 3\sec\theta\,\tan\theta d\theta$

Now we proceed to make substitutions and integrate

$$\int \frac{1}{\sqrt{x^2 - 9}}\, dx = \int \frac{1}{3\tan\theta}\, 3\sec\theta\,\tan\theta\, d\theta = \int \sec\theta d\theta = ln|\sec\theta + \tan\theta| + c$$

To return to the original variable, we proceed to construct the right triangle

As $x = 3\sec\theta$ you have: $\sec\theta = \frac{x}{3}$

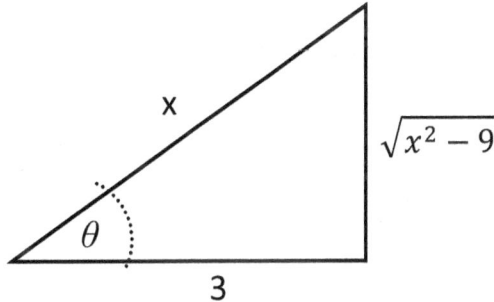

$$\int \frac{1}{\sqrt{x^2 - 9}}\, dx = ln|\sec\theta + \tan\theta| + c = ln\left|\frac{x}{3} + \frac{\sqrt{x^2 - 9}}{3}\right| + C$$

$$\int \frac{1}{\sqrt{x^2 - 9}}\, dx = ln\left|\frac{x + \sqrt{x^2 - 9}}{3}\right| + c = ln\left|x + \sqrt{x^2 - 9}\right| - ln3 + C$$

Note: As $(-ln3)$ es una constante se le suma a la constate (C)

$$\boxed{\int \frac{1}{\sqrt{x^2 - 9}}\, dx = ln\left|x + \sqrt{x^2 - 9}\right| + C}$$

094

$$\int \frac{\sqrt{1-x^2}}{x^4}\, dx$$

It is appreciated that the integral is of the form $\sqrt{a^2 - u^2}$

Where: $a = 1$, $x = sen\,\theta$, $dx = \cos\theta d\theta$, $\sqrt{1-x^2} = \cos\theta$,

Now we proceed to make substitutions and integrate

$$\int \frac{\sqrt{1-x^2}}{x^4}\, dx = \int \frac{\cos\theta \cos\theta\, d\theta}{sen^4\theta} = \int \frac{\cos^2\theta\, d\theta}{sen^2\theta\, sen^2\theta}$$

$$\int \frac{\sqrt{1-x^2}}{x^4}\, dx = \int \frac{\cos^2\theta}{sen^2\theta}\frac{1}{sen^2\theta}\, d\theta = \int cot^2\theta\, csc^2\theta\, d\theta$$

$$\int \frac{\sqrt{1-x^2}}{x^4}\, dx = \int u^2\, du = -\frac{1}{3}cot^3\theta + C$$

To return to the original variable, we proceed to construct the right triangle

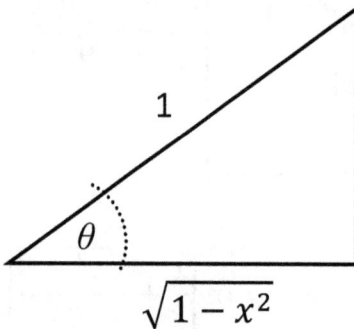

Como: $x = sen\,\theta \rightarrow sen\,\theta = \frac{x}{1}$

$$\int \frac{\sqrt{1-x^2}}{x^4}\, dx = -\frac{1}{3}cot^3\theta + C = -\frac{1}{3}\left(\frac{\sqrt{1-x^2}}{x}\right)^3 + C$$

$$\boxed{\int \frac{\sqrt{1-x^2}}{x^4}\, dx = -\frac{(1-x^2)^{\frac{3}{2}}}{3x^3} + C}$$

$$\int \frac{1}{x\,\sqrt{4x^2+9}}\,dx$$

095

The integral is rewritten to determine that the form is: $\sqrt{u^2+a^2}$

$$\int \frac{1}{x\,\sqrt{4x^2+9}}\,dx = \int \frac{1}{x\,\sqrt{(2x)^2+3^2}}\,dx$$

Where: $a = 3$, $\sqrt{4x^2+9} = 3\sec\theta$, $2x = 3\tan\theta$, $x = \frac{3}{2}\tan\theta$, $dx = \frac{3}{2}\sec^2\theta\,d\theta$

Now we proceed to make substitutions and integrate

$$\int \frac{1}{x\,\sqrt{4x^2+9}}\,dx = \int \frac{\frac{3}{2}\sec^2\theta\,d\theta}{\frac{3}{2}\tan\theta\,\,3\sec\theta} = \frac{1}{3}\int \frac{\sec\theta\,d\theta}{\tan\theta}$$

$$\int \frac{1}{x\,\sqrt{4x^2+9}}\,dx = \frac{1}{3}\int \frac{\frac{1}{\cos\theta}\,d\theta}{\frac{sen\theta}{\cos\theta}} = \frac{1}{3}\int \frac{\cos\theta\,d\theta}{sen\theta\,\cos\theta}$$

$$\int \frac{1}{x\,\sqrt{4x^2+9}}\,dx = \frac{1}{3}\int \csc\theta\,d\theta = -\frac{1}{3}\ln|\csc\theta+\cot\theta| + C$$

To return to the original variable, we proceed to construct the right triangle

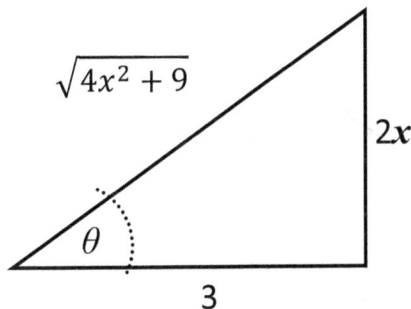

$$2x = 3\tan\theta$$
$$\tan\theta = \frac{2x}{3}$$

$$\int \frac{1}{x\sqrt{4x^2+9}}\,dx = -\frac{1}{3}\ln|\csc\theta+\cot\theta|+c = -\frac{1}{3}\ln\left|\frac{\sqrt{4x^2+9}}{2x}+\frac{3}{2x}\right|+C$$

$$\boxed{\int \frac{1}{x\sqrt{4x^2+9}}\,dx = -\frac{1}{3}\ln\left|\frac{\sqrt{4x^2+9}+3}{2x}\right|+C}$$

096

$$\int \frac{-5x}{(x^2+5)^{\frac{3}{2}}}\,dx$$

The integral is rewritten to determine that the form is: $\sqrt{u^2+a^2}$

$$\int \frac{-5x}{(x^2+5)^{\frac{3}{2}}}\,dx = \int \frac{-5x}{\left(\sqrt{x^2+(\sqrt5)^2}\right)^3}\,dx$$

Where: $a=\sqrt5$, $\sqrt{x^2+(\sqrt5)^2}=\sqrt{5^3}\sec^3\theta$, $x=\sqrt5\tan\theta$, $dx=\sqrt5\sec^2\theta d\theta$

$$\int \frac{-5x}{\left(\sqrt{x^2+(\sqrt5)^2}\right)^3}\,dx = \int \frac{-5\sqrt5\tan\theta}{\sqrt{5^3}\sec^3\theta}\,\sqrt5\sec^2\theta d\theta$$

$$\int \frac{-5x}{\left(\sqrt{x^2+(\sqrt5)^2}\right)^3}\,dx = -\frac{5}{\sqrt5}\int \frac{\tan\theta}{\sec\theta}\,d\theta = -\frac{5}{\sqrt5}\int \sin\theta\,d\theta = \frac{5}{\sqrt5}\cos\theta+C$$

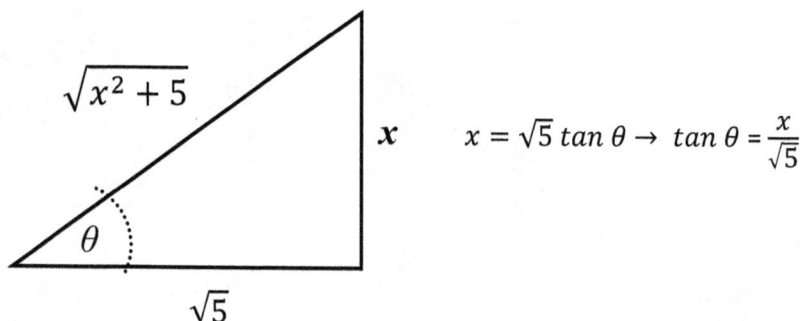

x $x=\sqrt5\tan\theta \rightarrow \tan\theta=\dfrac{x}{\sqrt5}$

$\sqrt{x^2+5}$

θ

$\sqrt5$

$$\int \frac{-5x}{(x^2+5)^{\frac{3}{2}}}\,dx = \frac{5}{\sqrt{5}}\cos\theta + C = \frac{5}{\sqrt{5}}\frac{\sqrt{5}}{\sqrt{x^2+5}} + C$$

$$\boxed{\int \frac{-5x}{(x^2+5)^{\frac{3}{2}}}\,dx = \frac{5\sqrt{x^2+5}}{x^2+5} + C}$$

$$\int e^{2x}\sqrt{1+e^{2x}}\,dx \qquad 097$$

The integral is rewritten to determine that the form is: $\sqrt{u^2+a^2}$

$$\int e^{2x}\sqrt{1+e^{2x}}\,dx = \int (e^x)^2\sqrt{1+(e^x)^2}\,dx$$

Where

$$a = 1, \sqrt{1+(e^x)^2} = \sec\theta, e^x = \tan\theta \rightarrow xlne = \ln(tan\theta) \rightarrow$$
$$\rightarrow x = \ln(tan\theta) \rightarrow dx = \frac{1}{tan\theta}(\sec^2\theta)d\theta \rightarrow dx = \frac{\sec^2\theta}{tan\theta}d\theta$$

Now we proceed to make substitutions and integrate

$$\int (e^x)^2\sqrt{1+(e^x)^2}\,dx = \int \tan^2\theta\,\sec\theta\frac{\sec^2\theta}{tan\theta}d\theta = \int \sec^3\theta\tan\theta\,d\theta$$

$$\int (e^x)^2\sqrt{1+(e^x)^2}\,dx = \int \sec^2\theta\,\sec\theta\,\tan\theta\,d\theta = \int u^2du = \frac{\sec^3\theta}{3} + C$$

To return to the original variable, we proceed to construct the right triangle

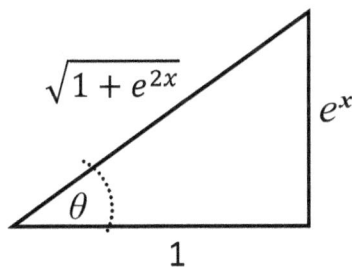

$$e^x = \tan\theta \rightarrow \tan\theta = \frac{e^x}{1}$$

$$\int e^{2x}\sqrt{1+e^{2x}}\ dx = \frac{1}{3}(\sqrt{1+e^{2x}})^3 + C = \frac{1}{3}(1+e^{2x})^{\frac{3}{2}} + C$$

$$\boxed{\int e^{2x}\sqrt{1+e^{2x}}\ dx = \frac{1}{3}(1+e^{2x})^{\frac{3}{2}} + C}$$

098 $\qquad\boxed{\int e^x\sqrt{1-e^{2x}}\ dx}$

The integral is rewritten to determine that the form is: $\sqrt{a^2 - u^2}$

$$\int e^x\sqrt{1-e^{2x}}\ dx = \int e^x\sqrt{1-(e^x)^2}\ dx$$

Where: $a = 1$, $\sqrt{1-(e^x)^2} = \cos\theta$, $e^x = sen\ \theta \rightarrow xlne = \ln(sen\theta) \rightarrow$

$$\rightarrow x = \ln(sen\theta) \rightarrow dx = \frac{1}{sen\theta}(cos\theta)d\theta \rightarrow dx = \frac{cos\theta}{sen\theta}d\theta$$

$$\int e^x\sqrt{1-(e^x)^2}\ dx = \int sen\theta\ cos\theta\ \frac{cos\theta}{sen\theta}d\theta = \int cos^2\theta\ d\theta$$

$$\int e^x\sqrt{1-(e^x)^2}\ dx = \int \frac{1+cos2\theta}{2}d\theta = \frac{1}{2}\int (1+cos2\theta)\ d\theta$$

$$\int e^x\sqrt{1-(e^x)^2}\ dx = \frac{1}{2}\int d\theta + \frac{1}{2}\int cos2\theta d\theta$$

$$\int e^x\sqrt{1-(e^x)^2}\ dx = \frac{1}{2}\theta + \frac{1}{4}sen2\theta + c = \frac{1}{2}\theta + \frac{1}{2}sen\theta cos\theta + c$$

To return to the original variable, we proceed to construct the right triangle

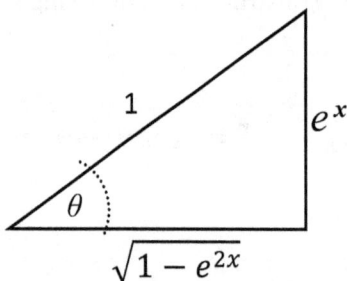

$$e^x = sen\ \theta \rightarrow sen\ \theta = \frac{e^x}{1}$$
$$\theta = \text{arcsen}\ (e^x)$$

$$\int e^x \sqrt{1 - e^{2x}}\ dx = \frac{1}{2}\arcsen(e^x) + \frac{1}{2}\ e^x\ \sqrt{1 - e^{2x}} + C$$

$$\boxed{\int e^x\sqrt{1 - e^{2x}}\ dx = \frac{1}{2}\left[\arcsen(e^x) + e^x\ \sqrt{1 - e^{2x}}\right] + C}$$

$$\boxed{\int \frac{1}{4 + 4x^2 + x^4}\ dx}$$

To rewrite the integral, the polynomial must be factored $4 + 4x^2 + x^4$

$x^4 + 4x^2 + 4 = (x^2)^2 + 4x^2 + 4$. Resulting in a perfect square trinomial and its factorization is $(x^2 + 2)^2$.

$$\int \frac{1}{4 + 4x^2 + x^4}\ dx = \int \frac{1}{(x^2 + 2)^2}\ dx = \int \frac{1}{(x^2 + (\sqrt{2})^2)^2}\ dx$$

Where: $a = \sqrt{2}$, $x = \sqrt{2}\ tan\theta$, $dx = \sqrt{2}\ sec^2\theta d\theta$

The form is $\sqrt{a^2 + u^2} = a\ sec\ \theta$. Both members of the equation are squared and he get:
$a^2 + u^2 = a^2 sec^2\theta \rightarrow x^2 + 2 = (\sqrt{2})^2\ sec^2\theta = 2sec^2\theta$.

$$\int \frac{1}{(x^2 + 2)^2}\ dx = \int \frac{\sqrt{2}\ sec^2\theta d\theta}{(2sec^2\theta)^2} = \int \frac{\sqrt{2}\ sec^2\theta d\theta}{4\ sec^4\theta} = \frac{\sqrt{2}}{4}\int \frac{d\theta}{sec^2\theta}$$

$$\int \frac{1}{(x^2 + 2)^2}\ dx = \frac{\sqrt{2}}{4}\int cos^2\theta d\theta = \frac{\sqrt{2}}{8}\int (1 + cos2\theta)d\theta = \frac{\sqrt{2}}{8}\int d\theta + \frac{\sqrt{2}}{8}\int (cos2\theta)d\theta$$

$$\int \frac{1}{(x^2 + 2)^2}\ dx = \frac{\sqrt{2}}{8}\theta + \frac{\sqrt{2}}{16}\ sen2\theta + c = \frac{\sqrt{2}}{8}\theta + \frac{\sqrt{2}}{8}\ sen\theta\ cos\theta + c$$

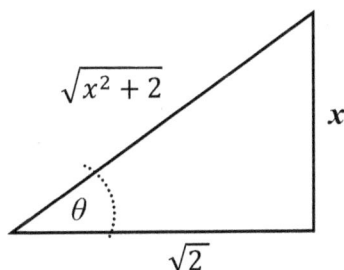

$$x = \sqrt{2}\ tan\theta \rightarrow tan\theta = \frac{x}{\sqrt{2}}$$

$$\theta = arctan\frac{x}{\sqrt{2}}$$

$$\int \frac{1}{4+4x^2+x^4}\, dx = \frac{\sqrt{2}}{8}\left(arctan\frac{x}{\sqrt{2}} + \frac{x\sqrt{2}}{x^2+2}\right) + C$$

$$\boxed{\int \frac{1}{4+4x^2+x^4}\, dx = \frac{1}{4}\left(\frac{x}{x^2+2} + \frac{\sqrt{2}}{2}arctan\frac{x}{\sqrt{2}}\right) + C}$$

100 $\qquad\qquad\qquad \boxed{\int arcsec2x\; dx}$

The integral is solved applying the technique of integration by parts.

$$u = arcsec2x \longrightarrow du = \frac{2\,dx}{2x\sqrt{4x^2-1}} \rightarrow du = \frac{dx}{x\sqrt{4x^2-1}}$$

$$dv = dx \rightarrow v = x$$

$$\int arcsec2x\; dx = x\,arcsec2x - \int \frac{dx}{\sqrt{4x^2-1}} \quad \textbf{Equation 1}$$

Now we proceed to solve the integral $\int \frac{dx}{\sqrt{4x^2+1}}$ by trigonometric substitution

$$\int \frac{dx}{\sqrt{4x^2-1}} = \int \frac{dx}{\sqrt{(2x)^2-1}}$$

Where: $a = 1$, $2x = \sec\theta$, $x = \frac{1}{2}\sec\theta$, $dx = \frac{1}{2}\sec\theta\tan\theta\, d\theta$, $\sqrt{(2x)^2-1} = \tan\theta$

$$\int \frac{dx}{\sqrt{(2x)^2-1}} = \int \frac{\frac{1}{2}\sec\theta\tan\theta\, d\theta}{\tan\theta} = \frac{1}{2}\int \sec\theta d\theta = \frac{1}{2}\,ln|\sec\theta+\tan\theta| + C$$

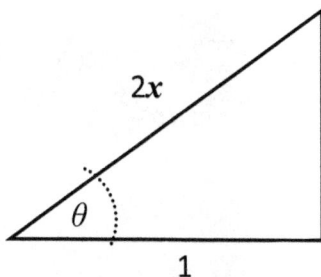

$$2x = \sec\theta \rightarrow \sec\theta = \frac{2x}{1}$$

$$\int \frac{dx}{\sqrt{4x^2 - 1}} = \frac{1}{2} \ln \left| 2x + \sqrt{4x^2 - 1} \right| + C$$

The previous result is replaced in equation 1 and you get

$$\int arcsec2x \ dx = x \ arcsec2x - \frac{1}{2} \ln \left| 2x + \sqrt{4x^2 - 1} \right| + C$$

$$\int \frac{x}{\sqrt{x^2 + 6x + 5}} \, dx$$

101

To transform the expression that is in the radical in one of the forms studied in the theoretical foundation, we proceed to complete squares.

$$x^2 + 6x + 5 = (x^2 + 6x + 9) + 5 - 9 = (x + 3)^2 - 4$$

$$\int \frac{x}{\sqrt{x^2 + 6x + 5}} \, dx = \int \frac{x}{\sqrt{(x + 3)^2 - 4}} \, dx = \int \frac{x}{\sqrt{(x + 3)^2 - 2^2}} \, dx$$

Where:

$$a = 2, x + 3 = 2sec\theta \rightarrow x = 2sec\theta - 3, \quad dx = 2sec\theta \ tan\theta \ d\theta, \sqrt{(x + 3)^2 - 4} = 2tan\theta$$

$$\int \frac{x}{\sqrt{(x + 3)^2 - 2^2}} \, dx = \int \frac{2sec\theta - 3}{2tan\theta} 2sec\theta \ tan\theta \ d\theta = \int (2sec^2\theta - 3 \ sec\theta) d\theta$$

$$\int \frac{x}{\sqrt{(x + 3)^2 - 2^2}} \, dx = 2 \int sec^2\theta \ d\theta - 3 \int sec\theta \ d\theta$$

$$\int \frac{x}{\sqrt{(x + 3)^2 - 2^2}} \, dx = 2tan\theta - 3ln|sec\theta + tan\theta| + C$$

$$\int \frac{x}{\sqrt{x^2 + 6x + 5}} \, dx = \sqrt{x^2 + 6x + 5} - 3ln \left| \frac{(x + 3) + \sqrt{x^2 + 6x + 5}}{2} \right| + C$$

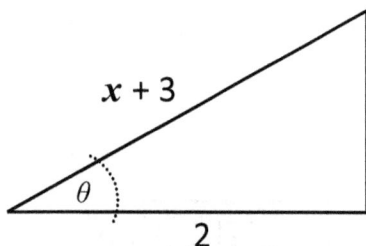

$$\sqrt{(x+3)^2 - 4} \quad x + 3 = 2sec\theta \;\rightarrow\; sec\theta = \frac{x+3}{2}$$

$$\int \frac{x}{\sqrt{x^2 + 6x + 5}}\, dx = \sqrt{x^2 + 6x + 5} - 3ln\left|\frac{(x+3) + \sqrt{x^2 + 6x + 5}}{2}\right| + C$$

102

$$\int \frac{\sqrt{1-x}}{\sqrt{x}}\, dx$$

To transform the numerator into the form $\sqrt{a^2 - u^2}$, becomes (x) in the form of power, as well: $(\sqrt{x})^2$. Y el 1 en 1^2.

$$\int \frac{\sqrt{1-x}}{\sqrt{x}}\, dx = \int \frac{\sqrt{1^2 - (\sqrt{x})^2}}{\sqrt{x}}\, dx$$

$a = 1,\ \sqrt{x} = sen\theta \;\rightarrow\; x = sen^2\theta \;\rightarrow\; dx = 2sen\theta\, cos\theta\, d\theta,\ \sqrt{1^2 - (\sqrt{x})^2} = cos\theta$

Now we proceed to make the substitutions and solve the integrals

$$\int \frac{\sqrt{1^2 - (\sqrt{x})^2}}{\sqrt{x}}\, dx = \int \frac{cos\theta}{sen\theta}\, 2sen\theta\, cos\theta\, d\theta = 2\int cos^2\theta\, d\theta$$

$$\int \frac{\sqrt{1^2 - (\sqrt{x})^2}}{\sqrt{x}}\, dx = \int (1 + cos2\theta)\, d\theta = \int d\theta + \int cos2\theta\, d\theta$$

$$\int \frac{\sqrt{1^2 - (\sqrt{x})^2}}{\sqrt{x}}\, dx = \theta + \frac{1}{2}\, sen2\theta + C = \theta + sen\theta\, cos\theta + C$$

To return to the original variable, we proceed to construct the right triangle.

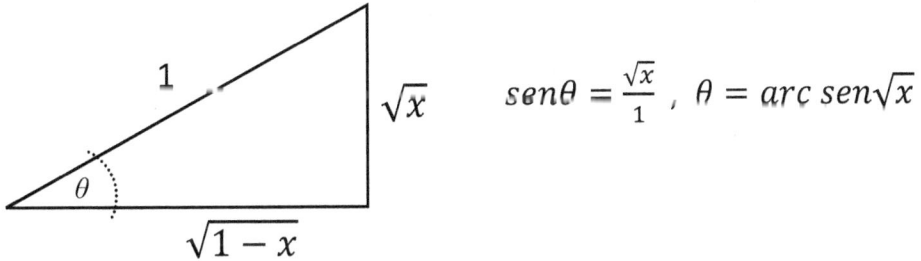

$$sen\theta = \frac{\sqrt{x}}{1} \; , \; \theta = arc\; sen\sqrt{x}$$

$$\boxed{\int \frac{\sqrt{1-x}}{\sqrt{x}}\; dx = arc\; sen\sqrt{x} + \sqrt{x}\sqrt{1-x} + C}$$

$$\boxed{\int \frac{1}{\sqrt{4x-x^2}}\; dx}$$

103

Complete squares similarly to problem number 101

$$4x - x^2 = -(x^2 - 4x) = -[(x^2 - 4x + 4) - 4]$$
$$4x - x^2 = 4 - (x-2)^2 = 2^2 - (x-2)^2$$

Therefore, the integral is rewritten as follows:

$$\int \frac{1}{\sqrt{4x-x^2}}\; dx = \int \frac{1}{\sqrt{2^2 - (x-2)^2}}\; dx$$

Where: $a = 2$, $x - 2 = 2sen\theta, x = 2sen\theta + 2, dx = 2cos\theta d\theta, \sqrt{2^2 - (x-2)^2} = 2cos\theta$.

Now we proceed to make the substitutions and solve the integral

$$\int \frac{1}{\sqrt{2^2 - (x-2)^2}}\; dx = \int \frac{2cos\theta d\theta}{2cos\theta} = \int d\theta = \theta + C$$

Then, it returns to the original variable in the following way

$$x - 2 = 2sen\theta \quad \rightarrow \quad sen\theta = \frac{x-2}{2} \quad \rightarrow \quad \theta = arcsen\frac{x-2}{2}$$

$$\int \frac{1}{\sqrt{4x - x^2}}\, dx = arcsen\left(\frac{x - 2}{2}\right) + C$$

$$\boxed{\int \frac{1}{\sqrt{4x - x^2}}\, \boldsymbol{dx} = \boldsymbol{arcsen}\left(\frac{x - 2}{2}\right) + \boldsymbol{C}}$$

104 $\qquad\qquad\qquad\qquad\qquad\qquad\qquad \boxed{\displaystyle\int (x + 1)\sqrt{x^2 + 2x + 2}\, dx}$

The procedure is similar to the previous problem.

$$x^2 + 2x + 2 = (x^2 + 2x + 1) + 2 - 1 = (x + 1)^2 + 1$$

Therefore, the integral is rewritten in this way

$$\int (x + 1)\sqrt{x^2 + 2x + 2}\, dx = \int (x + 1)\sqrt{(x + 1)^2 + 1}\, dx$$

Where: $a = 1$, $x + 1 = tan\theta$, $x = tan\theta - 1$, $dx = sec^2\theta d\theta$, $\sqrt{(x + 1)^2 + 1} = sec\,\theta$.

Then substitutions are made and the integral is solved

$$\int (x + 1)\sqrt{x^2 + 2x + 2}\, dx = \int (tan\theta - 1 + 1)sec\,\theta\, sec^2\theta d\theta$$

$$\int (x + 1)\sqrt{x^2 + 2x + 2}\, dx = \int sec^2\theta sec\theta tan\theta\, d\theta$$

$$\int (x + 1)\sqrt{x^2 + 2x + 2}\, dx = \int u^2 du = \frac{1}{3}sec^3\theta + C$$

Now to return to the original variable, it is done as follows

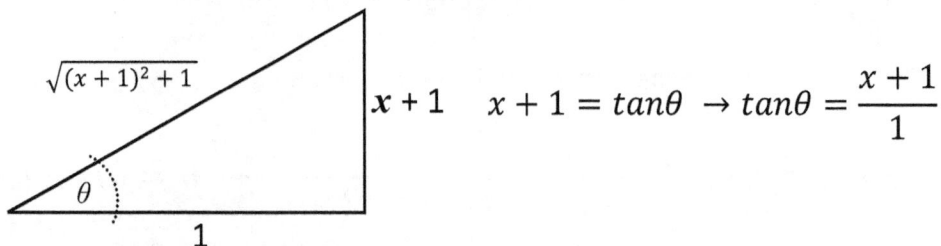

$x + 1 \qquad x + 1 = tan\theta \rightarrow tan\theta = \dfrac{x + 1}{1}$

$$\int (x+1)\sqrt{x^2+2x+2}\ dx = \frac{1}{3}\left(\sqrt{(x+1)^2+1}\right)^3 + C$$

$$\boxed{\int (x+1)\sqrt{x^2+2x+2}\ \boldsymbol{dx} = \frac{1}{3}(x^2+2x+2)^{\frac{3}{2}} + \boldsymbol{C}}$$

$$\boxed{\int \frac{t}{(1-t^2)^{\frac{3}{2}}}\ dt}$$

105

The integral is rewritten

$$\int \frac{t}{(1-t^2)^{\frac{3}{2}}}\ dt = \int \frac{t}{(\sqrt{1-t^2})^3}\ dt$$

Where $a=1$, $t = sen\theta$, $dt = cos\theta d\theta$, $\sqrt{1-t^2} = cos\theta$

Then substitutions are made and the integral is solved

$$\int \frac{t}{(\sqrt{1-t^2})^3}\ dt = \int \frac{sen\theta}{cos^3\theta}\ cos\theta d\theta = \int \frac{sen\theta}{cos^2\theta}\ d\theta$$

To solve the integral it is done $u = cos\theta$ y $du = -sen\theta d\theta$.

$$\int \frac{sen\theta}{cos^2\theta}\ d\theta = -\int \frac{du}{u^2} = -\int u^{-2}du = -\frac{u^{-1}}{-1} + C = \frac{1}{cos\theta} + C$$

Now to return to the original variable, it is done as follows

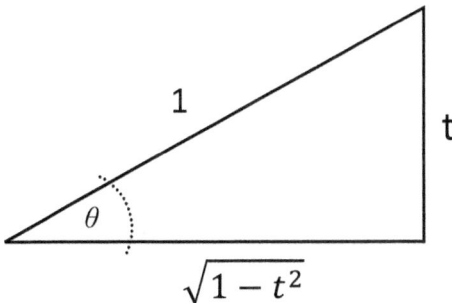

$$t = sen\theta \ \rightarrow \ sen\theta = \frac{t}{1}$$

$$\int \frac{t}{(1-t^2)^{\frac{3}{2}}} \, dt = \frac{1}{cos\theta} + C = \frac{1}{\sqrt{1-t^2}} + C$$

$$\boxed{\int \frac{t}{(1-t^2)^{\frac{3}{2}}} \, dt = \frac{1}{\sqrt{1-t^2}} + C}$$

106 $\qquad\qquad\qquad\qquad\qquad\qquad\qquad\boxed{\int x \, arcsenx dx}$

To solve the integral requires the application of the technique of integration by parts. (Technique studied in this book). The result you get is the following

$$\int x \, arcsenx dx = \frac{x^2}{2} \, arcsenx - \frac{1}{2}\int \frac{x^2}{\sqrt{1-x^2}} \, dx \;\; \textbf{Equation 1}$$

To solve the resulting integral $\int \frac{x^2}{\sqrt{1-x^2}} \, dx$, the technique under study is applied.

Where: $a = 1$, $x = sen\theta$, $dx = cos\theta d\theta$, $\sqrt{1-x^2} = cos\theta$

$$\int \frac{x^2}{\sqrt{1-x^2}} dx = \int \frac{sen^2\theta}{cos\theta} \, cos\theta d\theta = \int sen^2\theta \, d\theta = \frac{1}{2}\int (1-cos2\theta)d\theta$$

$$\int \frac{x^2}{\sqrt{1-x^2}} dx = \frac{1}{2}\int d\theta - \frac{1}{2}\int cos2\theta d\theta = \frac{1}{2}\theta - \frac{1}{4}sen2\theta + C = \frac{1}{2}\theta - \frac{1}{2}sen\theta cos\theta + C$$

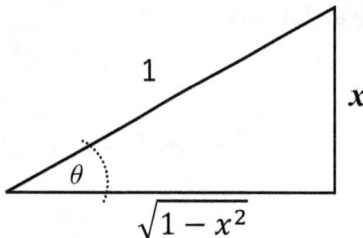

$x = sen\theta \quad \rightarrow \quad sen\theta = \frac{x}{1}$

$\theta = arcsenx$

$$\int \frac{x^2}{\sqrt{1-x^2}} dx = \frac{1}{2}\theta - \frac{1}{2}sen\theta cosx + C = \frac{1}{2}arcsenx - \frac{1}{2}x\sqrt{1-x^2} + C$$

This result is replaced in equation 1

$$\int x\, arcsenx dx = \frac{x^2}{2}\, arcsonx - \frac{1}{4}arcsenx + \frac{1}{4}x\sqrt{1-x^2} + C$$

$$\int x\, arcsenx dx = \frac{1}{4}\left[2x^2 arcsenx - arcsenx + x\sqrt{1-x^2}\right] + C$$

$$\boxed{\int x\, arcsenx dx = \frac{1}{4}\left[(2x^2-1)arcsenx + x\sqrt{1-x^2}\right] + C}$$

$$\boxed{\int \frac{x^3+x+1}{x^4+2x^2+1}dx}$$

107

The integral is separated into two integrals in the following way

$$\int \frac{x^3+x+1}{x^4+2x^2+1}dx = \int \frac{x^3+x}{x^4+2x^2+1}dx + \int \frac{1}{x^4+2x^2+1}dx \quad \textbf{Equation 1}$$

The first integral is multiplied and divided by four and resolved, the denominator of the second integral is factorized and resolved.

Resolution of the first integral.

$$\int \frac{x^3+x}{x^4+2x^2+1}dx = \frac{1}{4}\int \frac{4x^3+4x}{x^4+2x^2+1}dx$$

Where: $u = x^4+2x^2+1 \rightarrow du = (4x^3+4x)\,dx$

$$\frac{1}{4}\int \frac{4x^3+4x}{x^4+2x^2+1}dx = \frac{1}{4}\int \frac{du}{u} = \frac{1}{4}ln[x^4+2x^2+1] + C$$

$$\boxed{\frac{1}{4}\int \frac{4x^3+4x}{x^4+2x^2+1}dx = \frac{1}{2}ln[(x^2+1)] + C \quad \textbf{Equation 2}}$$

Resolution of the second integral

$$\int \frac{1}{x^4 + 2x^2 + 1}\, dx$$

The denominator is factored and the integral is rewritten.

$$x^4 + 2x^2 + 1 = (x^2)^2 + 2x^2 + 1 = (x^2 + 1)^2$$

$$\int \frac{1}{x^4 + 2x^2 + 1}\, dx = \int \frac{1}{(x^2 + 1)^2}\, dx$$

Where: $a = 1$, $x = tan\theta$, $dx = sec^2\theta d\theta$

The form is: $\sqrt{a^2 + u^2} = a\,sec\theta$. Then both members of the equation are squared and obtained.

$$a^2 + u^2 = a^2 sec^2\theta \quad \rightarrow \quad x^2 + 1^2 = sec^2\theta$$

$$\int \frac{1}{(x^2 + 1^2)^2}\, dx = \int \frac{sec^2\theta d\theta}{sec^4\theta} = \int \frac{1}{sec^2\theta}\, d\theta = \int cos^2\theta d\theta$$

To solve the resulting integral, the integral number 99 is taken as reference, therefore we have: $\int cos^2\theta d\theta = \frac{1}{2}(\theta + sen\theta\, cos\theta) + C$

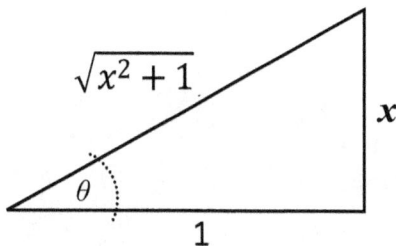

$$x = tan\theta \rightarrow tan\theta = \frac{x}{1}$$
$$\theta = arctanx$$

$$\int \frac{1}{(x^2 + 1)^2}\, dx = \frac{1}{2}(\theta + sen\theta\, cos\theta) + C = \frac{1}{2}(arctanx + \frac{x}{\sqrt{x^2 + 1}}\frac{1}{\sqrt{x^2 + 1}}) + C$$

$$\int \frac{1}{(x^2 + 1)^2}\, dx = \frac{1}{2}(\theta + sen\theta\, cos\theta) + C = \frac{1}{2}(arctanx + \frac{x}{x^2 + 1}) + C$$

$$\int \frac{1}{(x^2+1)^2}dx = \frac{1}{2}\left(arctanx + \frac{x}{(x^2+1)}\right) + C \ \textbf{\textit{Equation 3}}$$

Substitute equation 2 and equation 3 in equation 1

$$\int \frac{x^3+x+1}{x^4+2x^2+1}dx = \int \frac{x^3+x}{x^4+2x^2+1}dx + \int \frac{1}{x^4+2x^2+1}dx$$

$$\int \frac{x^3+x+1}{x^4+2x^2+1}dx = \frac{1}{2}ln[(x^2+1)] + \frac{1}{2}\left(arctanx + \frac{x}{(x^2+1)}\right) + C$$

$$\boxed{\int \frac{x^3+x+1}{x^4+2x^2+1}dx = \frac{1}{2}\left[ln(x^2+1) + \left(arctanx + \frac{x}{(x^2+1)}\right)\right] + C}$$

$$\boxed{\int \frac{\sqrt{4x^2+9}}{x^4}dx}$$ 108

The integral is rewritten.

$$\int \frac{\sqrt{4x^2+9}}{x^4}dx = \int \frac{\sqrt{(2x)^2+3^2}}{x^4}dx$$

Where: $a = 3$, $2x = 3tan\theta$, $x = \frac{3}{2}tan\theta$, $dx = \frac{3}{2}sec^2\theta d\theta$, $\sqrt{(2x)^2+3^2} = 3sec\,\theta$.

$$\int \frac{\sqrt{(2x)^2+3^2}}{x^4}dx = \int \frac{3sec\,\theta\,\frac{3}{2}sec^2\theta d\theta}{(\frac{3}{2}tan\theta)^4} = \int \frac{\frac{9}{2}sec^3\theta d\theta}{\frac{81}{16}tan^4\theta}$$

$$\int \frac{\sqrt{(2x)^2+3^2}}{x^4}dx = \frac{8}{9}\int \frac{cos\theta\,d\theta}{sen^4\theta} = \frac{8}{9}\int \frac{du}{u^4} = -\frac{8}{27}csc^3\theta + c$$

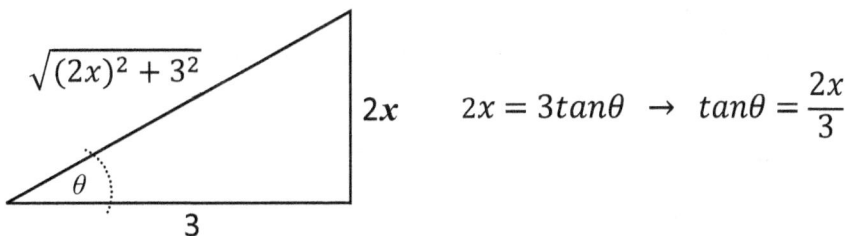

$2x = 3tan\theta \rightarrow tan\theta = \frac{2x}{3}$

$$\int \frac{\sqrt{4x^2 + 9}}{x^4} dx = -\frac{8}{27} csc^3\theta + C = -\frac{8}{27}\left(\frac{\sqrt{4x^2 + 9}}{2x}\right)^3 + C$$

$$\int \frac{\sqrt{4x^2 + 9}}{x^4} dx = -\frac{8}{27}\frac{\left(\sqrt{4x^2 + 9}\right)^3}{8x^3} + C = -\frac{\left(\sqrt{4x^2 + 9}\right)^3}{27x^3} + C$$

$$\boxed{\int \frac{\sqrt{4x^2 + 9}}{x^4} dx = -\frac{\left(\sqrt{4x^2 + 9}\right)^3}{27x^3} + C}$$

109

$$\int \frac{x}{\sqrt{x^2 + 4x + 8}} dx$$

Square is completed and the integral is rewritten.

$$x^2 + 4x + 8 = (x^2 + 4x + 4) + 8 - 4 = (x + 2)^2 + 2^2$$

$$\int \frac{x}{\sqrt{x^2 + 4x + 8}} dx = \int \frac{x}{\sqrt{(x + 2)^2 + 2^2}} dx$$

Where:

$$a = 2, x + 2 = 2tan\theta, x = 2tan\theta - 2, \qquad dx = 2sec^2\theta d\theta, \sqrt{(x + 2)^2 + 2^2} = 2sec\,\theta.$$

$$\int \frac{x}{\sqrt{(x + 2)^2 + 2^2}} dx = \int \frac{(2tan\theta - 2)2sec^2\theta d\theta}{2sec\,\theta} = \int (2tan\theta - 2)\,sec\theta d\theta$$

$$\int \frac{x}{\sqrt{(x + 2)^2 + 2^2}} dx = 2\int sec\theta\,tan\theta d\theta - 2\int sec\theta d\theta = 2sec\theta - 2ln[sec\theta + tan\theta] + C$$

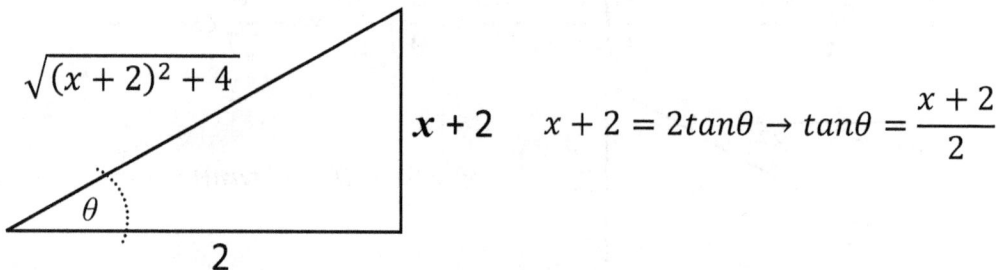

$\sqrt{(x + 2)^2 + 4}$

$x + 2$

$x + 2 = 2tan\theta \rightarrow tan\theta = \dfrac{x + 2}{2}$

θ

2

$$\int \frac{x}{\sqrt{x^2 + 4x + 8}}\, dx = 2\frac{\sqrt{(x+2)^2 + 4}}{2} - 2ln\left[\frac{\sqrt{(x+2)^2 + 4}}{2} + \frac{x+2}{2}\right] + C$$

$$\boxed{\int \frac{x}{\sqrt{x^2 + 4x + 8}}\, dx = \sqrt{x^2 + 4x + 8} - 2ln\left[\sqrt{x^2 + 4x + 8} + x + 2\right] + C}$$

$$\boxed{\int \frac{x^2}{\sqrt{2x - x^2}}\, dx} \qquad \boxed{110}$$

Square is completed and the integral is rewritten.

$$2x - x^2 = -(x^2 - 2x) = -[(x^2 - 2x + 1) - 1] = 1 - (x-1)^2$$

$$\int \frac{x^2}{\sqrt{2x - x^2}}\, dx = \int \frac{x^2}{\sqrt{1 - (x-1)^2}}\, dx$$

Where: $a = 1$, $x - 1 = sen\theta$, $x = sen\theta + 1$, $dx = cos\theta d\theta$, $\sqrt{1 - (x-1)^2} = cos\theta$

$$\int \frac{x^2}{\sqrt{1 - (x-1)^2}}\, dx = \int \frac{(sen\theta + 1)^2 cos\theta d\theta}{cos\theta} = \int (sen\theta + 1)^2\, d\theta$$

$$\int \frac{x^2}{\sqrt{1 - (x-1)^2}}\, dx = \int sen^2\theta\, d\theta + 2\int sen\theta\, d\theta + \int d\theta$$

$$\int \frac{x^2}{\sqrt{1 - (x-1)^2}}\, dx = \frac{1}{2}\int d\theta - \frac{1}{2}\int cos2\theta\, d\theta + 2\int sen\theta\, d\theta + \int d\theta$$

$$\int \frac{x^2}{\sqrt{1 - (x-1)^2}}\, dx = \frac{1}{2}\theta - \frac{1}{4}sen2\theta - 2cos\theta + \theta + C = \rightarrow$$

$$\rightarrow \frac{3}{2}\theta - \frac{1}{2}sen\theta cos\theta - 2cos\theta + C$$

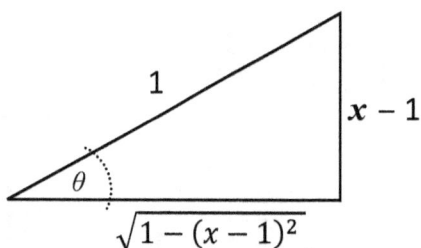

$$x - 1 = sen\theta \rightarrow sen\theta = \frac{x-1}{1}$$

$$\theta = arcsen(x-1)$$

$$\int \frac{x^2}{\sqrt{2x - x^2}}\, dx \rightarrow$$

$$\rightarrow = \frac{3}{2}arcsen(x-1) - \frac{1}{2}(x-1)\sqrt{1-(x-1)^2} - 2\sqrt{1-(x-1)^2} + C$$

$$\int \frac{x^2}{\sqrt{2x - x^2}}\, dx = \frac{3}{2}arcsen(x-1) - \frac{1}{2}\sqrt{2x - x^2}\,[x - 1 + 4] + C$$

$$\boxed{\int \frac{x^2}{\sqrt{2x - x^2}}\, dx = \frac{3}{2}arcsen(x-1) - \frac{1}{2}\sqrt{2x - x^2}\,[x + 3] + C}$$

111

$$\int \frac{1}{(x^2 + 3)^{\frac{3}{2}}}\, dx$$

The integral is rewritten as follows

$$\int \frac{1}{(x^2 + 3)^{\frac{3}{2}}}\, dx = \int \frac{1}{\left(\sqrt{x^2 + (\sqrt{3})^2}\right)^3}\, dx$$

Where: $a = \sqrt{3}$, $x = \sqrt{3}tan\theta$, $dx = \sqrt{3}sec^2\theta\, d\theta$, $\sqrt{x^2 + (\sqrt{3})^2} = \sqrt{3}sec\,\theta$

$$\int \frac{1}{\left(\sqrt{x^2 + (\sqrt{3})^2}\right)^3}\, dx = \int \frac{\sqrt{3}sec^2\theta\, d\theta}{(\sqrt{3}sec\,\theta)^3} = \int \frac{\sqrt{3}sec^2\theta\, d\theta}{\sqrt{27}sec^3\theta}$$

$$\int \frac{1}{\left(\sqrt{x^2+(\sqrt{3})^2}\right)^3}\,dx = \int \frac{\sqrt{3}\sec^2\theta d\theta}{\sqrt{27}\sec^3\theta} = \frac{1}{3}\int \frac{1}{\sec\theta}\,d\theta$$

$$\int \frac{1}{\left(\sqrt{x^2+(\sqrt{3})^2}\right)^3}\,dx = \frac{1}{3}\int \cos\theta\,d\theta = \frac{1}{3}\,sen\theta + C$$

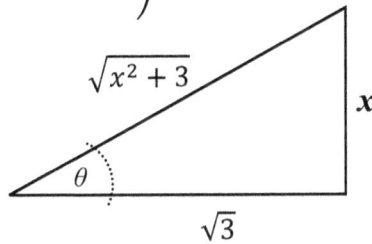

$$x = \sqrt{3}\tan\theta \;\rightarrow\; \tan\theta = \frac{x}{\sqrt{3}}$$

$$\boxed{\int \frac{1}{(x^2+3)^{\frac{3}{2}}}\,dx = \frac{1}{3}\frac{x}{\sqrt{x^2+3}} + C}$$

$$\boxed{\int \frac{1}{x\sqrt{4x^2+16}}\,dx}$$ **112**

The integral is rewritten in the following terms

$$\int \frac{1}{x\sqrt{4x^2+16}}\,dx = \int \frac{1}{x\sqrt{(2x)^2+4^2}}\,dx$$

Where: $a = 4$, $2x = 4\tan\theta$, $x = 2\tan\theta$, $dx = 2\sec^2\theta d\theta$, $\sqrt{(2x)^2+4^2} = 4\sec\theta$

$$\int \frac{1}{x\sqrt{(2x)^2+4^2}}\,dx = \int \frac{2\sec^2\theta d\theta}{2\tan\theta 4\sec\theta} = \frac{1}{4}\int \frac{\sec\theta d\theta}{\tan\theta}$$

$$\int \frac{1}{x\sqrt{(2x)^2+4^2}}\,dx = \frac{1}{4}\int \frac{\frac{1}{\cos\theta}}{\frac{sen\theta}{\cos\theta}}\,d\theta = \frac{1}{4}\int \frac{1}{sen\theta}\,d\theta = \frac{1}{4}\int \csc\theta d\theta$$

$$\int \frac{1}{x\sqrt{(2x)^2 + 4^2}}\, dx = -\frac{1}{4} ln[csc\theta + cot\theta] + C$$

Now to return to the original variable, it is done as follows

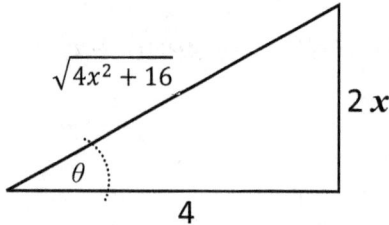

$$2x = 4tan\theta \rightarrow tan\theta = \frac{2x}{4}$$

$$\int \frac{1}{x\sqrt{(2x)^2 + 4^2}}\, dx = -\frac{1}{4} ln[csc\theta + cot\theta] + C = -\frac{1}{4} ln\left[\frac{\sqrt{4x^2 + 16}}{2x} + \frac{4}{2x} \right] + C$$

$$\boxed{\int \frac{1}{x\sqrt{4x^2 + 16}}\, dx = -\frac{1}{4} ln\left| \frac{4 + \sqrt{4x^2 + 16}}{2x} \right| + C}$$

113

$$\int x\sqrt{16 - 4x^2}\, dx$$

The integral is rewritten in the following terms

$$\int x\sqrt{16 - 4x^2}\, dx = \int x\sqrt{4^2 - (2x)^2}\, dx$$

Where: $a = 4$, $2x = 4sen\theta$, $x = 2sen\theta$, $dx = 2cos\theta d\theta$, $\sqrt{4^2 - (2x)^2} = 4cos\theta$

$$\int x\sqrt{4^2 - (2x)^2}\, dx = \int 2sen\theta\, 4cos\theta\, 2cos\theta\, d\theta$$

$$\int x\sqrt{4^2 - (2x)^2}\, dx = 16 \int cos^2\theta\, sen\theta\, d\theta$$

$$\int x\sqrt{4^2 - (2x)^2}\, dx = -16 \int u^2\, du = -\frac{16}{3} u^3 + C$$

$$\int x\sqrt{4^2 - (2x)^2}\, dx = -\frac{16}{3} \cos^3\theta + C$$

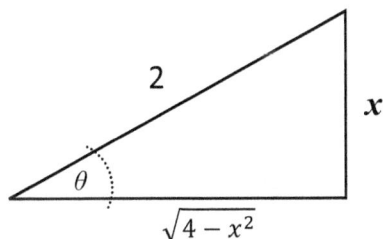

$$2x = 4 sen\theta \quad \rightarrow \quad sen\theta = \frac{x}{2}$$

$$\int x\sqrt{16 - 4x^2}\, dx = -\frac{16}{3} \cos^3\theta + C = -\frac{16}{3}\left(\frac{\sqrt{4 - x^2}}{2}\right)^3 + C$$

$$\int x\sqrt{16 - 4x^2}\, dx = -\frac{16}{24}\left(\sqrt{4 - x^2}\right)^3 + C$$

$$\boxed{\int x\sqrt{16 - 4x^2}\, dx = -\frac{2}{3}\left(\sqrt{4 - x^2}\right)^3 + C}$$

$$\boxed{\int \frac{x - 3}{(5 - 4x - x^2)^{\frac{3}{2}}}\, dx} \qquad \boxed{114}$$

The integral is rewritten and squares are completed

$$\int \frac{x - 3}{(5 - 4x - x^2)^{\frac{3}{2}}}\, dx = \int \frac{x - 3}{\left(\sqrt{5 - 4x - x^2}\right)^3}\, dx$$

$$5 - 4x - x^2 = -(x^2 + 4x - 5) = 3^2 - (x + 2)^2$$

$$\int \frac{x - 3}{(5 - 4x - x^2)^{\frac{3}{2}}}\, dx = \int \frac{x - 3}{\left(\sqrt{3^2 - (x + 2)^2}\right)^3}\, dx$$

$$a = 3, \quad x + 2 = 3 sen\theta, \quad x = 3 sen\theta - 2, \quad dx = 3 cos\theta d\theta,$$

$$\sqrt{3^2 - (x + 2)^2} = 3 cos\theta$$

149

$$\int \frac{x-3}{\left(\sqrt{3^2-(x+2)^2}\right)^3}\,dx = \int \frac{(3sen\theta-2-3)3cos\theta d\theta}{(3cos\theta)^3} = \int \frac{(3sen\theta-5)3cos\theta d\theta}{27cos^3\theta}$$

$$\int \frac{x-3}{\left(\sqrt{3^2-(x+2)^2}\right)^3}\,dx = \int \frac{9sen\theta\,cos\theta-15cos\theta d\theta}{27cos^3\theta} = \frac{9}{27}\int \frac{sen\theta cos\theta d\theta}{cos^3\theta} - \frac{15}{27}\int \frac{cos\theta d\theta}{cos^3\theta}$$

$$\int \frac{x-3}{\left(\sqrt{3^2-(x+2)^2}\right)^3}\,dx = \frac{1}{3}\int \frac{sen\theta d\theta}{cos^2\theta} - \frac{5}{9}\int \frac{d\theta}{cos^2\theta}$$

$$\int \frac{x-3}{\left(\sqrt{3^2-(x+2)^2}\right)^3}\,dx = \frac{1}{3}\int (sec\theta\,tan\theta)d\theta - \frac{5}{9}\int sec^2 d\theta$$

$$\int \frac{x-3}{\left(\sqrt{3^2-(x+2)^2}\right)^3}\,dx = \frac{1}{3}sec\theta - \frac{5}{9}tan\theta + C$$

To return to the original variable, the following rectangle triangle is constructed.

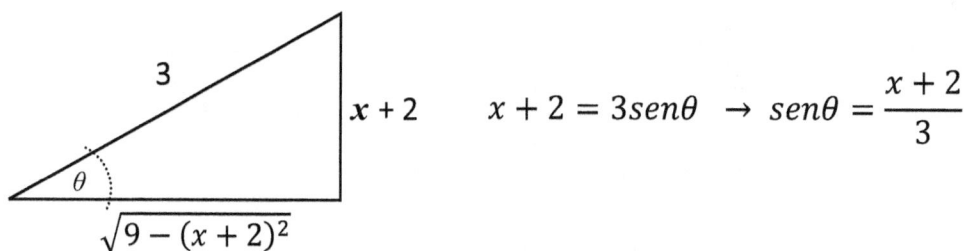

$$x+2 = 3sen\theta \quad \rightarrow \quad sen\theta = \frac{x+2}{3}$$

$$\int \frac{x-3}{(5-4x-x^2)^{\frac{3}{2}}}\,dx = \frac{1}{3}\frac{3}{\sqrt{9-(x+2)^2}} - \frac{5}{9}\frac{(x+2)}{\sqrt{9-(x+2)^2}} + C$$

$$\boxed{\int \frac{x-3}{(5-4x-x^2)^{\frac{3}{2}}}\,dx = \frac{-(5x+1)}{9\sqrt{9-(x+2)^2}} + C}$$

115

$$\int \frac{1}{\sqrt{25-x^2}}\,dx$$

$$\int \frac{1}{\sqrt{25-x^2}}\,dx = \int \frac{1}{\sqrt{5^2-x^2}}\,dx$$

Where: $a=5$, $x=5sen\theta$, $dx=5cos\theta d\theta$, $\sqrt{5^2-x^2}=5cos\theta$

$$\int \frac{1}{\sqrt{5^2-x^2}}\,dx = \int \frac{5cos\theta d\theta}{5cos\theta} = \int d\theta = \theta + C$$

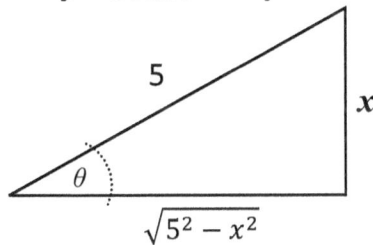

$x=5sen\theta \rightarrow sen\theta = \frac{x}{5}$

$\theta = arcsen\frac{x}{5}$

$$\int \frac{1}{\sqrt{25-x^2}}\,dx = \theta + C = arcsen\frac{x}{5}+C \rightarrow \boxed{\int \frac{1}{\sqrt{25-x^2}}\,dx = arcsen\frac{x}{5}+C}$$

116

$$\int \frac{x+3}{\sqrt{4x^2+4x-3}}\,dx$$

The integral is rewritten by completing squares

$$4x^2+4x-3 = 4\left(x^2+x-\frac{3}{4}\right)$$

$$4x^2+4x-3 = 4\left[(x^2+x+\frac{1}{4})-\frac{3}{4}-\frac{1}{4}\right]$$

$$4x^2+4x-3 = 4\left[\left(x+\frac{1}{2}\right)^2-\frac{4}{4}\right]$$

$$4x^2+4x-3 = 4\left[\left(x+\frac{1}{2}\right)^2-1\right]$$

$$\int \frac{x+3}{\sqrt{4x^2+4x-3}}\,dx = \int \frac{x+3}{\sqrt{4\left[\left(x+\frac{1}{2}\right)^2-1\right]}}\,dx = \frac{1}{2}\int \frac{x+3}{\sqrt{\left[\left(x+\frac{1}{2}\right)^2-1\right]}}\,dx$$

Where: $a=1$, $x+\frac{1}{2}=sec\theta$, $x=sec\theta-\frac{1}{2} \rightarrow dx=sec\theta\,tan\theta\,d\theta$, $\sqrt{\left[\left(x+\frac{1}{2}\right)^2-1\right]}=tan\theta$

$$\frac{1}{2}\int \frac{x+3}{\sqrt{\left[\left(x+\frac{1}{2}\right)^2-1\right]}}\,dx = \frac{1}{2}\int \frac{\left(sec\theta-\frac{1}{2}+3\right)sec\theta\,tan\theta\,d\theta}{tan\theta}$$

$$\frac{1}{2}\int \frac{x+3}{\sqrt{\left[\left(x+\frac{1}{2}\right)^2-1\right]}} = \frac{1}{2}\int \frac{\left(sec\theta+\frac{5}{2}\right)sec\theta\,tan\theta\,d\theta}{tan\theta}$$

$$\frac{1}{2}\int \frac{x+3}{\sqrt{\left[\left(x+\frac{1}{2}\right)^2-1\right]}}\,dx = \frac{1}{2}\int sec^2\theta\,d\theta + \frac{5}{4}\int sec\,\theta\,d\theta$$

$$\frac{1}{2}\int \frac{x+3}{\sqrt{\left[\left(x+\frac{1}{2}\right)^2-1\right]}}\,dx = \frac{1}{2}\,tan\theta + \frac{5}{4}\,ln|sec\theta+tan\theta|+C$$

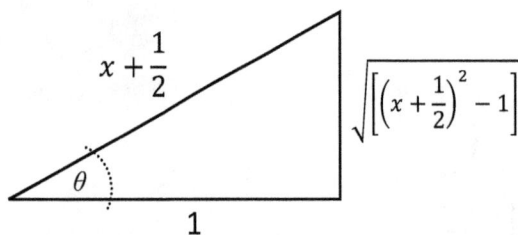

$x+\frac{1}{2}$ $\sqrt{\left[\left(x+\frac{1}{2}\right)^2-1\right]}$ $x+\frac{1}{2}=sec\theta \rightarrow sec\theta=\dfrac{x+\frac{1}{2}}{1}$

θ

1

$$\int \frac{x+3}{\sqrt{4x^2+4x-3}}\,dx = \frac{1}{2}\,tan\theta + \frac{5}{4}\,ln|sec\theta+tan\theta|+C$$

$$\int \frac{x+3}{\sqrt{4x^2+4x-3}}dx = \frac{1}{2}\sqrt{\left(x+\frac{1}{2}\right)^2-1} + \frac{5}{4}ln\left|x+\frac{1}{2}+\sqrt{\left(x+\frac{1}{2}\right)^2-1}\right| + C$$

$$\frac{1}{2}\int \frac{x+3}{\sqrt{\left[\left(x+\frac{1}{2}\right)^2-1\right]}}dx = \frac{1}{2}\sqrt{x^2+x-\frac{3}{4}} + \frac{5}{4}ln\left|\frac{2x+1}{2}+\sqrt{x^2+x-\frac{3}{4}}\right| + C$$

$$\frac{1}{2}\int \frac{x+3}{\sqrt{\left[\left(x+\frac{1}{2}\right)^2-1\right]}}dx = \frac{1}{2}\sqrt{x^2+x-\frac{3}{4}} + \frac{5}{4}ln\left|\frac{(2x+1)+2\sqrt{x^2+x-\frac{3}{4}}}{2}\right| + C$$

$$\frac{1}{2}\int \frac{x+3}{\sqrt{\left[\left(x+\frac{1}{2}\right)^2-1\right]}}dx = \frac{1}{4}\sqrt{4x^2+4x-3} + \frac{5}{4}ln\left|\frac{(2x+1)+2\sqrt{4x^2+4x-3}}{2}\right| + C$$

$$\boxed{\int \frac{x+3}{\sqrt{4x^2+4x-3}}dx = \frac{1}{4}\sqrt{4x^2+4x-3} + \frac{5}{4}ln\left|(2x+1)+2\sqrt{4x^2+4x-3}\right| + C}$$

$$\boxed{\int \frac{x^2}{(9-x^2)^{\frac{3}{2}}}dx}$$

117

The integral is rewritten

$$\int \frac{x^2}{(9-x^2)^{\frac{3}{2}}}dx = \int \frac{x^2}{\left(\sqrt{3^2-x^2}\right)^3}dx$$

Where: $a=3$, $x=3sen\theta$, $dx=3cos\theta d\theta$, $\sqrt{3^2-x^2}=3cos\theta$

$$\int \frac{x^2}{\left(\sqrt{3^2 - x^2}\right)^3}\,dx = \int \frac{(3sen\theta)^2}{(3cos\theta)^3}\,3cos\theta d\theta = \int \frac{9sen^2\theta}{27cos^3\theta}\,3cos\theta d\theta$$

$$\int \frac{x^2}{\left(\sqrt{3^2 - x^2}\right)^3}\,dx = \int \frac{sen^2\theta}{cos^2\theta}\,d\theta = \int tan^2\theta d\theta = \int (sec^2\theta - 1)d\theta$$

$$\int \frac{x^2}{\left(\sqrt{3^2 - x^2}\right)^3}\,dx = \int sec^2\theta d\theta - \int d\theta = tan\theta - \theta + C$$

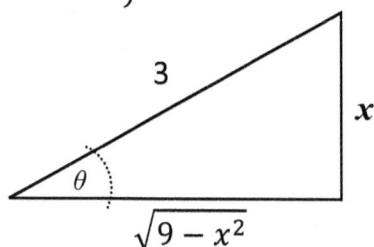

$$x = 3sen\theta \;\rightarrow\; sen\theta = \frac{x}{3}$$
$$\theta = arcsen\frac{x}{3}$$

$$\boxed{\int \frac{x^2}{(9 - x^2)^{\frac{3}{2}}}\,dx = \frac{x}{\sqrt{9 - x^2}} - arcsen\frac{x}{3} + C}$$

118

$$\int (9 - x^2)^{-\frac{3}{2}}\,dx$$

The integral is rewritten

$$\int (9 - x^2)^{-\frac{3}{2}}\,dx = \int \frac{1}{(9 - x^2)^{\frac{3}{2}}}\,dx = \int \frac{1}{\left(\sqrt{3^2 - x^2}\right)^3}\,dx$$

Where: $a = 3$, $x = 3sen\theta$, $dx = 3cos\theta d\theta$, $\sqrt{3^2 - x^2} = 3cos\theta$

$$\int \frac{1}{\left(\sqrt{3^2 - x^2}\right)^3}\,dx = \int \frac{3cos\theta d\theta}{(3cos\theta)^3} = \frac{1}{9}\int \frac{d\theta}{cos^2\theta}$$

$$\int \frac{1}{\left(\sqrt{3^2 - x^2}\right)^3}\,dx = \frac{1}{9}\int sec^2\theta d\theta = \frac{1}{9}tan\theta + C$$

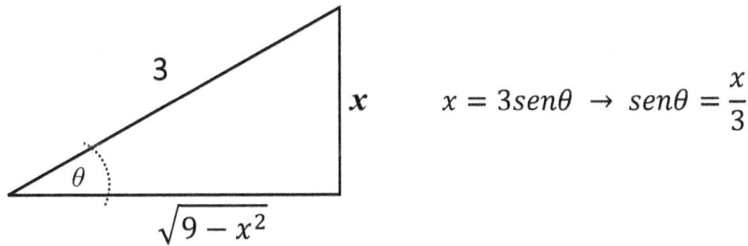

$$x = 3sen\theta \rightarrow sen\theta = \frac{x}{3}$$

$$\int (9-x^2)^{-\frac{3}{2}} dx = \frac{1}{9}tan\theta + C = \frac{x}{9\sqrt{9-x^2}} + C$$

$$\boxed{\int (9-x^2)^{-\frac{3}{2}} dx = \frac{x}{9\sqrt{9-x^2}} + C}$$

$$\int \frac{x^2 dx}{\sqrt{21+4x-x^2}}$$

119

The square of the polynomial is completed

$$21 + 4x - x^2 = -(x^2 - 4x - 21) = -(x^2 - 4x + 4) - 21 - 4$$

$$21 + 4x - x^2 = -[(x-2)^2 - 25] = 25 - (x-2)^2 = 5^2 - (x-2)^2$$

$$\int \frac{x^2 dx}{\sqrt{21+4x-x^2}} = \int \frac{x^2 dx}{\sqrt{5^2 - (x-2)^2}}$$

$$a = 5, \ x - 2 = 5sen\theta, \ x = 5sen\theta + 2, \ dx = 5cos\theta d\theta, \ \sqrt{5^2 - (x-2)^2} = 5cos\theta$$

$$\int \frac{x^2 dx}{\sqrt{5^2 - (x-2)^2}} = \int \frac{(5sen\theta + 2)^2 5cos\theta d\theta}{5cos\theta}$$

$$\int \frac{x^2 dx}{\sqrt{5^2 - (x-2)^2}} = \int (5sen\theta + 2)^2 d\theta$$

$$\int \frac{x^2 dx}{\sqrt{5^2 - (x-2)^2}} = \int (25 sen^2\theta + 20 sen\theta + 4)\, d\theta$$

$$\int \frac{x^2 dx}{\sqrt{5^2 - (x-2)^2}} = 25 \int sen^2\theta\, d\theta + 20 \int sen\theta\, d\theta + 4 \int d\theta$$

$$\int \frac{x^2 dx}{\sqrt{5^2 - (x-2)^2}} = \frac{25}{2} \int (1 - cos2\theta)\, d\theta + 20 \int sen\theta\, d\theta + 4 \int d\theta$$

$$\int \frac{x^2 dx}{\sqrt{5^2 - (x-2)^2}} = \frac{25}{2} \int d\theta - \frac{25}{2} \int cos2\theta\, d\theta + 20 \int sen\theta\, d\theta + 4 \int d\theta$$

$$\int \frac{x^2 dx}{\sqrt{5^2 - (x-2)^2}} = \frac{25}{2}\theta - \frac{25}{4} sen2\theta - 20 cos\theta + 4\theta + C$$

$$\int \frac{x^2 dx}{\sqrt{5^2 - (x-2)^2}} = \frac{33}{2}\theta - \frac{25}{2} sen\theta cos\theta - 20 cos\theta + C$$

Remember: $sen^2\theta = \frac{(1 - cos2\theta)}{2}$	Remember: $sen2\theta = 2 sen\theta cos\theta$

$$\int \frac{x^2 dx}{\sqrt{5^2 - (x-2)^2}} = \frac{33}{2}\theta - \frac{25}{2} sen\theta cos\theta - 20 cos\theta + C$$

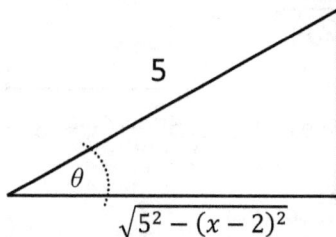

$$x - 2 = 5 sen\theta \rightarrow sen\theta = \frac{x-2}{5}$$

$$\theta = arcsen\frac{x-2}{5}$$

$$\int \frac{x^2 dx}{\sqrt{21 + 4x - x^2}} \rightarrow$$

$$\rightarrow = \frac{33}{2} arcsen\frac{x-2}{5} - \frac{25}{2}\frac{(x-2)}{5}\frac{\sqrt{25 - (x-2)^2}}{5} - 20\frac{\sqrt{25 - (x-2)^2}}{5} + C$$

$$\int \frac{x^2dx}{\sqrt{21+4x-x^2}} = \frac{33}{2}arcsen\frac{x-2}{5} - \frac{1}{2}(x-2)\sqrt{21+4x-x^2} - 4\sqrt{21+4x-x^2} + C$$

$$\int \frac{x^2dx}{\sqrt{21+4x-x^2}} = \frac{33}{2}arcsen\frac{x-2}{5} - \sqrt{21+4x-x^2}\left[\frac{1}{2}(x-2)+4\right] + C$$

$$\int \frac{x^2dx}{\sqrt{21+4x-x^2}} = \frac{33}{2}arcsen\frac{x-2}{5} - \sqrt{21+4x-x^2}\left(\frac{x+6}{2}\right) + C$$

$$\boxed{\int \frac{x^2dx}{\sqrt{21+4x-x^2}} = \frac{33}{2}arcsen\frac{x-2}{5} - \sqrt{21+4x-x^2}\left(\frac{x+6}{2}\right) + C}$$

$$\boxed{\int \frac{dx}{\sqrt{2-5x^2}}} \qquad \text{120}$$

The integral is rewritten

$$\int \frac{dx}{\sqrt{2-5x^2}} = \int \frac{dx}{\sqrt{(\sqrt{2})^2-(\sqrt{5}x)^2}}$$

$$a=\sqrt{2}, \sqrt{5}\,x=\sqrt{2}sen\theta, x=\frac{\sqrt{2}}{\sqrt{5}}sen\theta, dx=\frac{\sqrt{2}}{\sqrt{5}}cos\theta d\theta, \quad \sqrt{(\sqrt{2})^2-(\sqrt{5}x)^2}=\sqrt{2}cos\theta$$

$$\int \frac{dx}{\sqrt{2-5x^2}} = \int \frac{dx}{\sqrt{(\sqrt{2})^2-(\sqrt{5}x)^2}} = \int \frac{\frac{\sqrt{2}}{\sqrt{5}}cos\theta d\theta}{\sqrt{2}cos\theta}$$

$$\int \frac{dx}{\sqrt{2-5x^2}} = \int \frac{dx}{\sqrt{(\sqrt{2})^2-(\sqrt{5}x)^2}} = \frac{1}{\sqrt{5}}\int d\theta = \frac{\sqrt{5}}{5}\theta + C$$

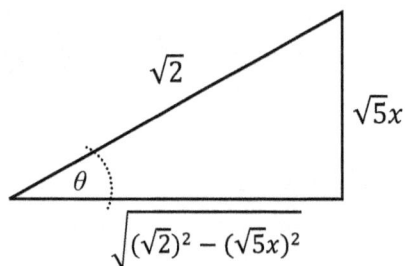

$$\sqrt{5}\,x = \sqrt{2}\,sen\theta \rightarrow sen\theta = \frac{\sqrt{5}x}{\sqrt{2}}$$

$$\theta = arcsen\frac{\sqrt{5}x}{\sqrt{2}}$$

$$\int \frac{dx}{\sqrt{2-5x^2}} = \frac{\sqrt{5}}{5}\ arcsen\frac{\sqrt{5}x}{\sqrt{2}} = \frac{\sqrt{5}}{5}\ arcsen\sqrt{\frac{5}{2}}\,x + C$$

$$\boxed{\int \frac{dx}{\sqrt{2-5x^2}} = \frac{\sqrt{5}}{5}\ arcsen\sqrt{\frac{5}{2}}\,x + C}$$

In the following exercises are the integrals of rational functions, applying the method of integration of partial fractions

$$\int \frac{1}{x^2 - 1}\, dx$$

|2|

The denominator is factored to find the factors. (It is suggested to the reader to study previously the theoretical foundation of this book).

To factor a difference of squares

$$x^2 - 1 = (x + 1)(x - 1) \quad \leftarrow \quad a^2 - b^2 = (a + b)(a - b)$$

The factors are linear and distinct. A simple fraction must be included for each factor

$$\frac{1}{x^2 - 1} = \frac{1}{(x + 1)(x - 1)} = \frac{A}{(x + 1)} + \frac{B}{(x - 1)}$$

Where A and B are numerical values to be calculated. In order to perform the calculation, denominators are eliminated,

$$\frac{1}{(x + 1)(x - 1)} = \frac{A}{(x + 1)} + \frac{B}{(x - 1)} \rightarrow 1 = A(x - 1) + B(x + 1)$$

The above equation is known as the fundamental equation.

$$1 = A(x - 1) + B(x + 1) \rightarrow 1 = Ax - A + Bx + B$$

By grouping the terms in x and the independent terms we have.

$$1 = (A + B)x + (-A + B)$$

Since there are no terms in x before equality, A + B = 0. And the independent terms are equal to 1, -A + B = 1 Consequently, the following system of equations is obtained:

$$A + B = 0 \quad Equation\ 1$$
$$-A + B = 1 \quad Equation\ 2$$

When solving the system of equations, the following numerical values are obtained for A, B.

$$\begin{array}{rl} A + B = 0 & \textit{Equation 1} \\ -A + B = 1 & \textit{Equation 2} \\ \hline 2B = 1 & \end{array}$$

Therefore, we have: $2B = 1 \;\rightarrow\; B = \frac{1}{2}$

Replace the value of B in equation 1 or in equation 2:

$$A + B = 0 \;\rightarrow\; A + \frac{1}{2} = 0 \;\rightarrow\; A = -\frac{1}{2}$$

Now we proceed to rewrite the integral as follows

$$\int \frac{1}{x^2 - 1}\,dx = \int \frac{1}{(x+1)(x-1)}\,dx = \int \left(\frac{A}{(x+1)} + \frac{B}{(x-1)}\right)dx$$

$$\int \frac{1}{x^2 - 1}\,dx = \int \frac{-\frac{1}{2}}{(x+1)}\,dx + \int \frac{\frac{1}{2}}{(x-1)}\,dx = -\frac{1}{2}\int \frac{dx}{(x+1)} + \frac{1}{2}\int \frac{dx}{(x-1)}$$

$$\int \frac{1}{x^2 - 1}\,dx = -\frac{1}{2}\ln|x+1| + \frac{1}{2}\ln|x-1| + C$$

$$\int \frac{1}{x^2 - 1}\,dx = \frac{1}{2}\ln\left|\frac{(x-1)}{(x+1)}\right| + C \quad (Logarithm\ properties\ applied)$$

$$\boxed{\int \frac{1}{x^2 - 1}\,dx = \frac{1}{2}\ln\left|\frac{(x-1)}{(x+1)}\right| + C}$$

Comment: Another procedure to calculate the values of A and B, is to substitute convenient values for x in the fundamental equation to obtain zeros in the other terms. That is to say.
$1 = A(x - 1) + B(x + 1)$ Fundamental equation.

$Para\ x = 1:\; 1 = A(1-1) + B(1+1) \rightarrow 1 = 2B \;\rightarrow\; B = \frac{1}{2}$

$Para\ x = -1:\; 1 = A(-1-1) + B(-1+1) \rightarrow 1 = -2A \rightarrow A = -\frac{1}{2}$

$$\int \frac{3}{x^2 + x - 2} dx$$

122

The procedure is analogous to the previous case.

$$x^2 + x - 2 = (x + 2)(x - 1) \leftarrow \quad \text{It is a case of factoring the form:}$$
$$x^2 + bx + c$$

The factors are linear and distinct. A simple fraction must be included for each factor.

$$\frac{3}{x^2 + x - 2} = \frac{3}{(x + 2)(x - 1)} = \frac{A}{(x + 2)} + \frac{B}{(x - 1)}$$

Eliminating denominators and solving the parentheses you have

$$3 = A(x - 1) + B(x + 2) \quad \rightarrow \quad 3 = Ax - A + Bx + 2B$$

Grouping the terms in X and the independent ones:

$$3 = (A + B)x + (-A + 2B)$$

The system of equations is constructed and solving A and B are obtained

$$
\begin{array}{ll}
A + B = 0 & \text{Equation 1} \\
-A + 2B = 3 & \text{Equation 2} \\
\hline
3B = 3 &
\end{array}
\qquad
\begin{array}{l}
3B = 3 \quad \rightarrow \quad \boldsymbol{B = 1} \\
A + B = 0 \quad \rightarrow \quad A + 1 = 0 \quad \rightarrow \quad \boldsymbol{A = -1}
\end{array}
$$

$$\int \frac{3}{x^2 + x - 2} dx = \int \frac{3}{(x + 2)(x - 1)} dx = \int \left[\frac{A}{(x + 2)} + \frac{B}{(x - 1)} \right] dx$$

$$\int \frac{3}{x^2 + x - 2} dx = -\int \frac{1}{(x + 2)} dx + \int \frac{1}{(x - 1)} dx = -ln[(x + 2)] + ln|(x - 1)| + C$$

$$\boxed{\int \frac{3}{x^2 + x - 2} dx = ln \left| \frac{(x - 1)}{(x + 2)} \right| + C}$$

Comment: For the calculation of A and B with the alternate procedure assume
$x = 1 \ y \ x = -2.$ ¡Do it!

123

$$\int \frac{5-x}{2x^2+x-1}\,dx$$

The denominator trinomial is of the form ax^2+bx+c. For its factorization we proceed to multiply the terms of the trinomial by the coefficient of x^2. Then the corresponding numbers are searched.

$$2x^2+x-1 = 4x^2+2(x)-2 = (2x)^2+(2x)-2 = (2x+2)(2x-1)$$

As the trinomial multiplied by 2 it is now divided by 2.

$$\frac{(2x+2)(2x-1)}{2} = (2x-1)(x+1) \quad\rightarrow\quad 2x^2+x-1 = (2x-1)(x+1)$$

The factors are linear and distinct. A simple fraction must be included for each factor.

$$\frac{5-x}{2x^2+x-1} = \frac{5-x}{(2x-1)(x+1)} = \frac{A}{(2x-1)} + \frac{B}{(x+1)}$$

Eliminating denominators and solving the parentheses we have:

$$5-x = A(x+1)+B(2x-1) \rightarrow 5-x = Ax+A+2Bx-B$$

Grouping the terms in x and independent: $5-x = (A+2B)x+(A-B)$

The system of equations is constructed and solving A and B are obtained

$$
\begin{array}{ll}
A+2B=-1 & \text{Equation 1} \\
A-B=5 & \text{Equation 2}
\end{array}
\qquad\rightarrow\qquad
\begin{array}{l}
A+2B=-1 \\
2A-2B=10 \\
\hline
3A \quad\;\; =9
\end{array}
$$

Despejando: Para A se tiene: $3A=9 \rightarrow \boldsymbol{A=3}$. Para B: $A-B=5 \rightarrow \boldsymbol{B=-2}$

$$\int \frac{5-x}{2x^2+x-1}\,dx = \int \frac{5-x}{(2x-1)(x+1)}\,dx = \int\left[\frac{3}{(2x-1)} + \frac{-2}{(x+1)}\right]dx$$

$$\int \frac{5-x}{2x^2+x-1}\,dx = 3\int \frac{dx}{(2x-1)} - 2\int \frac{dx}{(x-1)}$$

$$\boxed{\int \frac{5-x}{2x^2+x-1}\,dx = \frac{3}{2}ln|2x-1| - 2ln|x-1| + C}$$

$$\int \frac{x^2+12x+12}{x^3-4x}\,dx$$

124

Proceeding analogously to the previous cases, they have:

$$x^3 - 4x = x(x^2 - 4) = x(x+2)(x-2)$$

In factoring there are three linear and different factors. It is assigned to each factor a simple fraction.

$$\frac{x^2+12x+12}{x^3-4x} = \frac{x^2+12x+12}{x(x+2)(x-2)} = \frac{A}{x}+\frac{B}{(x+2)}+\frac{C}{(x-2)}$$

$$x^2+12x+12 = A(x+2)(x-2) + Bx(x-2) + Cx(x+2)$$

$$x^2+12x+12 = Ax^2 - 2Ax + 2Ax - 4A + Bx^2 - 2Bx + Cx^2 + 2Cx$$

$$x^2+12x+12 = (A+B+C)x^2 + (-2B+2C)x + (-4A)$$

Based on the previous step, the system of equations is constructed and solved.

$$
\begin{array}{ll}
A + B + \ C = 1 & \text{Equation 1} \\
{-2B + 2C} = 12 & \text{Equation 2} \\
-4A = 12 & \text{Equation 3}
\end{array}
$$

From equation 3 we obtain A: $-4A = 12 \ \rightarrow \ \boldsymbol{A = -3}$. The value is substituted in equation 1. $\ -3 + B + C = 1 \ \rightarrow \ B + C = 4$. Then multiply by 2 and combine with equation 2, resulting in the value of $C = 5$. With the values of A and C the value of B = -1 is obtained. Consequently, you have

$$\int \frac{x^2+12x+12}{x^3-4x}\,dx = \int \frac{x^2+12x+12}{x(x+2)(x-2)}\,dx = \int \left[\frac{A}{x}+\frac{B}{(x+2)}+\frac{C}{(x-2)}\right]dx$$

$$\int \frac{x^2+12x+12}{x^3-4x}\,dx = -3\int \frac{dx}{x} - \int \frac{dx}{(x+2)} + 5\int \frac{dx}{(x-2)}$$

$$\boxed{\int \frac{x^2+12x+12}{x^3-4x}\,dx = 5\ln|x-2| - \ln|x+2| - 3\ln|x| + C}$$

125

$$\int \frac{2x^3 - 4x^2 - 15x + 5}{x^2 - 2x - 8} dx$$

Since the power of the numerator is greater than the power of the denominator, the division of polynomials is previously carried out. The result is

$$\frac{2x^3 - 4x^2 - 15x + 5}{x^2 - 2x - 8} = 2x + \frac{x + 5}{x^2 - 2x - 8}$$

Then we proceed to factorize the polynomial of the divisor of the resulting fraction

$$x^2 - 2x - 8 = (x - 4)(x + 2)$$

$$\frac{2x^3 - 4x^2 - 15x + 5}{x^2 - 2x - 8} = 2x + \frac{x + 5}{(x - 4)(x + 2)}$$

Now the fraction is broken down into its simple fractions

$$\frac{x + 5}{(x - 4)(x + 2)} = \frac{A}{(x - 4)} + \frac{B}{(x + 2)}$$

Eliminating denominators and solving the parentheses we have:

$$x + 5 = A(x + 2) + B(x - 4) = Ax + 2A + Bx - 4B$$

$$x + 5 = (A + B)x + (2A - 4B)$$

The system of equations is constructed and it is solved to find the values of A and B.

$$
\begin{array}{ll}
A + B = 1 & \text{Equation 1} \\
2A - 4B = 5 & \text{Equation 2}
\end{array}
\quad \rightarrow \quad
\begin{array}{l}
-2A - 2B = -2 \\
2A - 4B = 5 \\
\hline
-6B = 3 \rightarrow B = -\frac{1}{2}
\end{array}
$$

For A you have: $A + B = 1 \rightarrow A - \frac{1}{2} = 1 \rightarrow \boldsymbol{A = \frac{3}{2}}$

$$\int \frac{2x^3 - 4x^2 - 15x + 5}{x^2 - 2x - 8} dx = 2 \int x dx + \frac{3}{2} \int \frac{dx}{(x - 4)} - \frac{1}{2} \int \frac{dx}{(x + 2)}$$

$$\int \frac{2x^3 - 4x^2 - 15x + 5}{x^2 - 2x - 8} dx = x^2 + \frac{3}{2} \ln|x - 4| - \frac{1}{2} \ln|x + 2| + C$$

$$\boxed{\int \frac{2x^3 - 4x^2 - 15x + 5}{x^2 - 2x - 8} dx = x^2 + \frac{3}{2} \ln|x - 4| - \frac{1}{2} \ln|x + 2| + C}$$

$$\int \frac{4x^2 + 2x - 1}{x^3 + x^2} dx$$

126

The polynomial of the denominator is factored.

$$x^3 + x^2 = x^2(x + 1)$$

Factoring has a quadratic factor and a linear factor. It is assigned to each factor its respective simple fraction.

$$\frac{4x^2 + 2x - 1}{x^3 + x^2} = \frac{4x^2 + 2x - 1}{x^2(x + 1)} = \frac{A}{x} + \frac{B}{x^2} + \frac{C}{(x + 1)}$$

Then denominators are eliminated, the parentheses are solved and the terms are grouped into x and independent terms.

$$4x^2 + 2x - 1 = Ax(x + 1) + B(x + 1) + Cx^2$$

$$4x^2 + 2x - 1 = Ax^2 + Ax + Bx + B + cx^2$$

$$4x^2 + 2x - 1 = (A + C)x^2 + (A + B)x + B$$

Now the system of equations is constructed and it is solved to find A, B and C.

$$
\begin{array}{lll}
A \quad + C = 4 & \quad Equation\ 1 \\
A + B \quad = 2 & \quad Equation\ 2 \\
\quad B \quad = -1 & \quad Equation\ 3 \\
\end{array}
$$

From equation 3 we obtain the value of B: B = -1
From equation 2 we obtain the value of A: A + B = 2 → A = 3
From equation 1 we obtain the value of C: A + C = 4 → C = 1
Consequently, you get:

$$\int \frac{4x^2 + 2x - 1}{x^3 + x^2} dx = \int \frac{4x^2 + 2x - 1}{x^2(x + 1)} dx = \int \left[\frac{A}{x} + \frac{B}{x^2} + \frac{C}{(x + 1)} \right] dx$$

$$\int \frac{4x^2 + 2x - 1}{x^3 + x^2} dx = 3 \int \frac{dx}{x} - \int \frac{dx}{x^2} + \int \frac{dx}{(x + 1)}$$

$$\boxed{\int \frac{4x^2 + 2x - 1}{x^3 + x^2} dx = 3ln|x| + \frac{1}{x} + ln|(x + 1)| + C}$$

127

$$\int \frac{x^2 + 3x - 4}{x^3 - 4x^2 + 4x}\,dx$$

The polynomial of the denominator is factored.

$$x^3 - 4x^2 + 4x = x(x^2 - 4x + 4) = x[(x-2)(x-2)] = x[(x-2)^2]$$

Factoring has a linear factor and a repeated factor

$$\frac{x^2 + 3x - 4}{x^3 - 4x^2 + 4x} = \frac{x^2 + 3x - 4}{x(x^2 - 4x + 4)} = \frac{x^2 + 3x - 4}{x[(x-2)^2\,]} = \frac{A}{x} + \frac{B}{(x-2)} + \frac{C}{(x-2)^2}$$

Then denominators are eliminated, the parentheses are solved and the terms are grouped into x and independent terms.

$$x^2 + 3x - 4 = A(x-2)(x-2) + Bx(x-2) + Cx$$

$$x^2 + 3x - 4 = Ax^2 - 4Ax + 4A + Bx^2 - 2Bx + Cx$$

$$x^2 + 3x - 4 = (A + B)x^2 + (-4A - 2B + C)x + (4A)$$

Now the system of equations is constructed and it is solved to find A, B and C.

$$
\begin{array}{rcll}
A + B & = & 1 & \textit{Equation 1} \\
-4A - 2B + C & = & 3 & \textit{Equation 2} \\
4A & = & -4 & \textit{Equation 3}
\end{array}
$$

From equation 3 we have: $A = -1$
From equation 1 we have that: $A + B = 1 \rightarrow B = 2$
From equation 2 we have: $-4A - 2B + C = 3 \rightarrow C = 3$

$$\int \frac{x^2 + 3x - 4}{x^3 - 4x^2 + 4x}\,dx = \int \left[\frac{A}{x} + \frac{B}{(x-2)} + \frac{C}{(x-2)^2}\right] dx$$

$$\int \frac{x^2 + 3x - 4}{x^3 - 4x^2 + 4x}\,dx = -\int \frac{dx}{x} + 2\int \frac{dx}{(x-2)} + 3\int \frac{dx}{(x-2)^2}$$

$$\boxed{\int \frac{x^2 + 3x - 4}{x^3 - 4x^2 + 4x}\,dx = -ln|x| + 2ln|x-2| - \frac{3}{x-2} + C}$$

$$\int \frac{x^2 - 1}{x^3 + x} dx$$

128

The denominator is factored.

$$x^3 + x = x(x^2 + 1)$$

Factoring has a quadratic factor and a linear factor. It is assigned to each factor its respective simple fraction.

$$\frac{x^2 - 1}{x^3 + x} = \frac{x^2 - 1}{x(x^2 + 1)} = \frac{A}{x} + \frac{Bx + c}{(x^2 + 1)}$$

Then denominators are eliminated, the parentheses are resolved and the terms are grouped into X and independent terms.

$$x^2 - 1 = A(x^2 + 1) + (Bx + C)(x)$$

$$x^2 - 1 = Ax^2 + A + Bx^2 + Cx = (A + B)x^2 + Cx + A$$

Now the system of equations is constructed and it is solved to find A, B and C.

$$
\begin{array}{lll}
A + B & = 1 & \textit{Equation 1} \\
C = 0 & & \textit{Equation 2} \\
A & = -1 & \textit{Equation 3}
\end{array}
$$

By simple inspection it is determined: $A = -1$, $B = 2$, $C = 0$

$$\int \frac{x^2 - 1}{x^3 + x} dx = \int \left[\frac{A}{x} + + \frac{Bx + C}{(x^2 + 1)}\right] dx = -\int \frac{dx}{x} + 2\int \frac{xdx}{(x^2 + 1)}$$

$$\int \frac{x^2 - 1}{x^3 + x} dx = -ln|x| + ln|x^2 + 1| + C$$

$$\boxed{\int \frac{x^2 - 1}{x^3 + x} dx = ln\left|\frac{x^2 + 1}{x}\right| + C}$$

129

$$\int \frac{x^2}{x^4 - 2x^2 - 8} dx$$

To factor the polynomial of the denominator first it is rewritten

$$x^4 - 2x^2 - 8 = (x^2)^2 - 2(x^2) - 8$$

The reader may appreciate that the result refers to a case of factoring the form $x^2 + bx + c$. To develop this case we proceed to find two numbers that multiplied den - 8 and that subtracted den - 2.

$$(x^2)^2 - 2(x^2) - 8 = (x^2 - 4)(x^2 + 2)$$

When solving the square difference of the first parenthesis the factorization is

$$x^4 - 2x^2 - 8 = (x - 2)(x + 2)(x^2 + 2)$$

Consequently, you have

$$\int \frac{x^2}{x^4 - 2x^2 - 8} dx = \int \frac{x^2}{(x - 2)(x + 2)(x^2 + 2)} dx$$

Then,

$$\frac{x^2}{(x - 2)(x + 2)(x^2 + 2)} = \frac{A}{(x - 2)} + \frac{B}{(x + 2)} + \frac{Cx + D}{(x^2 + 2)}$$

$$x^2 = A(x + 2)(x^2 + 2) + B(x - 2)(x^2 + 2) + (Cx + D)(x + 2)(x - 2)$$

$$x^2 = Ax^3 + 2Ax + 2Ax^2 + 4A + Bx^3 + 2Bx - 2Bx^2 - 4B + Cx^3 - 4Cx + Dx^2 - 4D$$

$$x^2 = (A + B + C)x^3 + (2A - 2B + D)x^2 + (2A + 2B - 4C)x + (4A - 4B - 4D)$$

$$
\begin{array}{llll}
A + & B & + C & = 0 & \text{Equation 1} \\
2A & - 2B & + D & = 1 & \text{Equation 2} \\
2A & + 2B - 4C & & = 0 & \text{Equation 3} \\
4A & - 4B & - 4D & = 0 & \text{Equation 4}
\end{array}
$$

Solving the system of equations we have the values for A, B, C and D.

$$A = \frac{1}{6} \ , \quad B = -\frac{1}{6} \ , \quad C = 0 \quad D = \frac{1}{3}$$

$$\int \frac{x^2}{x^4 - 2x^2 - 8} dx = \int \left[\frac{A}{(x-2)} + \frac{B}{(x+2)} + \frac{Cx+D}{(x^2+2)} \right] dx$$

$$\int \frac{x^2}{x^4 - 2x^2 - 8} dx = \int \frac{A}{(x-2)} dx + \int \frac{B}{(x+2)} dx + \int \frac{Cx+D}{(x^2+2)} dx$$

$$\int \frac{x^2}{x^4 - 2x^2 - 8} dx = \int \frac{\frac{1}{6}}{(x-2)} dx + \int \frac{-\frac{1}{6}}{(x+2)} dx + \int \frac{\frac{1}{3}}{(x^2+2)} dx$$

$$\int \frac{x^2}{x^4 - 2x^2 - 8} dx = \frac{1}{6}\int \frac{dx}{(x-2)} - \frac{1}{6}\int \frac{dx}{(x+2)} + \frac{1}{3}\int \frac{dx}{(x^2+2)}$$

$$\int \frac{x^2}{x^4 - 2x^2 - 8} dx = \frac{1}{6}\left[\int \frac{dx}{(x-2)} - \int \frac{dx}{(x+2)} + 2\int \frac{dx}{(x^2+2)} \right]$$

$$\int \frac{x^2}{x^4 - 2x^2 - 8} dx = \frac{1}{6}\left[ln|(x-2)| - ln|(x+2)| + \sqrt{2}arctan\frac{x}{\sqrt{2}} \right] + C$$

$$\int \frac{x^2}{x^4 - 2x^2 - 8} dx = \frac{1}{6}\left[ln\frac{(x-2)}{(x+2)} + \sqrt{2}arctan\left(\frac{x}{\sqrt{2}}\right) \right] + C$$

$$\boxed{\int \frac{x^2}{x^4 - 2x^2 - 8} dx = \frac{1}{6}\left[ln\frac{(x-2)}{(x+2)} + \sqrt{2}arctan\left(\frac{x}{\sqrt{2}}\right) \right] + C}$$

130

$$\int \frac{x}{16x^4 - 1} dx$$

To factor the polynomial of the denominator first it is rewritten

$$16x^4 - 1 = (4x^2)^2 - 1$$

The reader may notice that a difference of squares is established in the result. Its factorization is: $a^2 - b^2 = (a + b)(a - b)$. Therefore, we have:

$$(4x^2)^2 - 1 = (4x^2 - 1)(4x^2 + 1)$$

In the first parenthesis there is still a difference of squares which is factored:

$$(4x^2 - 1)(4x^2 + 1) = (2x - 1)(2x + 1)(4x^2 + 1)$$

As a result, you get

$$\int \frac{x}{16x^4 - 1} dx = \int \frac{x}{(2x - 1)(2x + 1)(4x^2 + 1)} dx$$

Then,

$$\frac{x}{(2x - 1)(2x + 1)(4x^2 + 1)} = \frac{A}{(2x - 1)} + \frac{B}{(2x + 1)} + \frac{Cx + D}{(4x^2 + 1)}$$

$$x = A(2x + 1)(4x^2 + 1) + B(2x - 1)(4x^2 + 1) + (Cx + D)(2x - 1)(2x + 1)$$

$$x = 8Ax^3 + 2Ax + 4Ax^2 + A + 8Bx^3 + 2Bx - 4Bx^2 - B + 4Cx^2 - Cx + 4Dx^2 - D$$

$$x = (8A + 8B)x^3 + (4A - 4B + 4C + 4D)x^2 + (2A + 2B - C)x + (A - B - D)$$

Now the system of equations is constructed and resolved

$$
\begin{array}{llllll}
8A & + 8B & & & = 0 & \textit{Equation 1} \\
4A & - 4B & + 4C & + 4D & = 0 & \textit{Equation 2} \\
2A & + 2B & - C & & = 1 & \textit{Equation 3} \\
A & - B & & - D & = 0 & \textit{Equation 4}
\end{array}
$$

Solving the system of equations you have the values for A, B, C and D.

$$A = \frac{1}{8} \, , \quad B = \frac{1}{8} \, , \quad C = -\frac{1}{2} \quad D = 0$$

$$\int \frac{x}{16x^4 - 1} dx = \int \left[\frac{A}{(2x - 1)} + \frac{B}{(2x + 1)} + \frac{Cx + D}{(4x^2 + 1)} \right] dx$$

$$\int \frac{x}{16x^4 - 1} dx = \int \frac{A}{(2x - 1)} dx + \int \frac{B}{(2x + 1)} dx + \int \frac{Cx + D}{(4x^2 + 1)} dx$$

$$\int \frac{x}{16x^4 - 1} dx = \int \frac{\frac{1}{8}}{(2x - 1)} dx + \int \frac{\frac{1}{8}}{(2x + 1)} dx + \int \frac{-\frac{1}{2}x}{(4x^2 + 1)} dx$$

$$\int \frac{x}{16x^4 - 1} dx = \frac{1}{8} \int \frac{dx}{(2x - 1)} + \frac{1}{8} \int \frac{dx}{(2x + 1)} - \frac{1}{2} \int \frac{xdx}{(4x^2 + 1)}$$

$$\int \frac{x}{16x^4 - 1} dx = \frac{1}{16} \int \frac{du}{u} + \frac{1}{16} \int \frac{dw}{w} - \frac{1}{16} \int \frac{dz}{z}$$

$$\int \frac{x}{16x^4 - 1} dx = \frac{1}{16} [\ln(2x - 1) + \ln(2x + 1) - \ln(4x^2 + 1)] + C$$

$$\int \frac{x}{16x^4 - 1} dx = \frac{1}{16} [\ln[(2x - 1)(2x + 1)] - \ln(4x^2 + 1)] + C$$

$$\int \frac{x}{16x^4 - 1} dx = \frac{1}{16} [ln(4x^2 - 1) - \ln(4x^2 + 1)] + C$$

$$\boxed{\int \frac{x}{16x^4 - 1} dx = \frac{1}{16} \ln \left| \frac{(4x^2 - 1)}{(4x^2 + 1)} \right| + C}$$

131

$$\int \frac{x^2 + 5}{x^3 - x^2 + x + 13} \, dx$$

When applying the Ruffini method, the root of the denominator polynomial is determined and is equal to -1. Therefore, its factorization is

$$x^3 - x^2 + x + 13 = (x + 1)(x^2 - 2x + 3)$$

Notice that the quadratic function is irreducible

$$\frac{x^2 + 5}{x^3 - x^2 + x + 13} = \frac{x^2 + 5}{(x + 1)(x^2 - 2x + 3)} = \frac{A}{(x + 1)} + \frac{Bx + C}{(x^2 - 2x + 3)}$$

$$x^2 + 5 = A(x^2 - 2x + 3) + (Bx + C)(x + 1)$$

$$x^2 + 5 = Ax^2 - 2Ax + 3A + Bx^2 + Bx + Cx + C$$

$$x^2 + 5 = (A + B)x^2 + (-2A + B + C)x + (3A + C)$$

Now we proceed to build the system of equations

$$
\begin{array}{rcll}
A + B & = 1 & \text{\textit{Equation 1}} \\
-2A + B + C & = 0 & \text{\textit{Equation 2}} \\
3A + C & = 5 & \text{\textit{Equation 3}}
\end{array}
$$

System solution $A = 1$ $\rightarrow$ $B = 0$ $\rightarrow$ $C = 2$

$$\int \frac{x^2 + 5}{x^3 - x^2 + x + 13} \, dx = \int \left[\frac{A}{(x + 1)} + \frac{Bx + C}{(x^2 - 2x + 3)} \right] dx$$

$$\int \frac{x^2 + 5}{x^3 - x^2 + x + 13} \, dx = \int \frac{dx}{(x + 1)} + 2 \int \frac{dx}{(x^2 - 2x + 3)}$$

$$\int \frac{x^2 + 5}{x^3 - x^2 + x + 13} \, dx = \int \frac{dx}{(x + 1)} + 2 \int \frac{dx}{(x - 1)^2 + (\sqrt{2})^2}$$

$$\int \frac{x^2 + 5}{x^3 - x^2 + x + 13} dx = ln|x + 1| + \frac{2}{\sqrt{2}} arctan\left(\frac{x-1}{\sqrt{2}}\right) + C$$

$$\boxed{\int \frac{x^2 + 5}{x^3 - x^2 + x + 13} dx = ln|x + 1| + \sqrt{2}\, arctan\left(\frac{x-1}{\sqrt{2}}\right) + C}$$

$$\boxed{\int \frac{1}{4x^2 - 9} dx} \qquad\qquad 132$$

The divisor polynomial is written and then factored

$$4x^2 - 9 = (2x)^2 - 3^2 = (2x - 3)(2x + 3)$$

$$\frac{1}{4x^2 - 9} = \frac{1}{(2x - 3)(2x + 3)} = \frac{A}{(2x - 3)} + \frac{B}{(2x + 3)}$$

$$1 = A(2x + 3) + B(2x - 3)$$

$$1 = 2Ax + 3A + 2Bx - 3B$$

$$1 = (2A + 2B)x + (3A - 3B)$$

Now the system of equations is constructed

$$\begin{array}{rll} 2A + \ 2B = 0 & \textit{Equation 1} \\ 3A - \ 3B = 1 & \textit{Equation 2} \end{array}$$

System solution $A = \frac{1}{6}$, $B = -\frac{1}{6}$

$$\int \frac{1}{4x^2 - 9} dx = \int \frac{1}{(2x - 3)(2x + 3)} dx = \int \left[\frac{A}{(2x - 3)} + \frac{B}{(2x + 3)}\right] dx$$

$$\int \frac{1}{4x^2 - 9} dx = \frac{1}{6}\int \frac{dx}{(2x - 3)} - \frac{1}{6}\int \frac{dx}{(2x + 3)}$$

$$\int \frac{1}{4x^2 - 9} dx = \frac{1}{12} ln|2x - 3| - \frac{1}{12} ln|2x + 3| + C$$

$$\boxed{\int \frac{1}{4x^2 - 9} dx = \frac{1}{12} ln \left|\frac{2x - 3}{2x + 3}\right| + C}$$

133 $\qquad\qquad \boxed{\int \frac{5x^2 - 12x - 12}{x^3 - 4x} dx}$

The polynomial of the divisor is factored

$$x^3 - 4x = x(x^2 - 4) = x(x - 2)(x + 2)$$

$$\frac{5x^2 - 12x - 12}{x^3 - 4x} = \frac{5x^2 - 12x - 12}{x(x - 2)(x + 2)} = \frac{A}{x} + \frac{B}{x - 2} + \frac{C}{x + 2}$$

$$5x^2 - 12x - 12 = A(x - 2)(x + 2) + Bx(x + 2) + Cx(x - 2)$$

$$5x^2 - 12x - 12 = (A + B + C)x^2 + (2B - 2C)x + (-4A)$$

Now we proceed to build the system of equations and its resolution

$$
\begin{array}{rcl}
A + B + C & = & 5 \quad \text{Equation 1} \qquad A = 3 \,, \; B = -2 \; y \; C = 4 \\
2B - 2C & = & -12 \quad \text{Equation 2} \\
-4A & = & 12 \quad \text{Equation 3}
\end{array}
$$

$$\int \frac{5x^2 - 12x - 12}{x^3 - 4x} dx = 3\int \frac{dx}{x} - 2\int \frac{dx}{x - 2} + 4\int \frac{dx}{x + 2}$$

$$\boxed{\int \frac{5x^2 - 12x - 12}{x^3 - 4x} dx = 3ln|x| - 2ln|x - 2| + 4ln|x + 2| + C}$$

$$\int \frac{x+1}{x^2+4x+3}\,dx$$

| 134

The polynomial of the divisor is factored $x^2 + 4x + 3 = (x+3)(x+1)$

$$\int \frac{x+1}{x^2+4x+3}\,dx = \int \frac{(x+1)}{(x+3)(x+1)}\,dx = \int \frac{dx}{(x+3)}$$

$$\boxed{\int \frac{x+1}{x^2+4x+3}\,dx = ln|x+3| + C}$$

$$\int \frac{x^3-x+3}{x^2+x-2}\,dx$$

| 135

Since the degree of the polynomial of the dividend is greater than the degree of the polynomial of the divisor, then the division of the polynomials is effected first. The result is

$$\frac{x^3-x+3}{x^2+x-2} = x - 1 + \frac{2x+1}{x^2+x-2}$$

Then we proceed to factorize the polynomial of the divisor of the resulting fraction

$$x^2 + x - 2 = (x+2)(x-1)$$

Now the fraction is broken down into its simple fractions:

$$\frac{2x+1}{x^2+x-2} = \frac{2x+1}{(x+2)(x-1)} = \frac{A}{x+2} + \frac{B}{x-1}$$

Eliminating denominators and solving the parentheses we have:

$$2x + 1 = A(x-1) + B(x+2)$$

$$2x + 1 = Ax - A + Bx + 2B$$

$$2x + 1 = (A + B)x + (-A + 2B)$$

The system of equations is constructed and it is solved to find the values of A and B

$$
\begin{array}{ll}
A + B = 2 & \text{Equation 1} \\
-A + 2B = 1 & \text{Equation 2}
\end{array}
\qquad \rightarrow \quad A = 1 \ y \ B = 1
$$

$$\int \frac{x^3 - x + 3}{x^2 + x - 2}\,dx = \int x\,dx - \int dx + \int \frac{dx}{x + 2} + \int \frac{dx}{x - 1}$$

$$\int \frac{x^3 - x + 3}{x^2 + x - 2}\,dx = \frac{1}{2}x^2 - x + \ln|x + 2| + \ln|x - 1| + C$$

$$\int \frac{x^3 - x + 3}{x^2 + x - 2}\,dx = \frac{1}{2}x^2 - x + \ln|(x + 2)(x - 1)| + C$$

$$\boxed{\int \frac{x^3 - x + 3}{x^2 + x - 2}\,dx = \frac{1}{2}x^2 - x + \ln|x^2 + x - 2| + C}$$

136

$$\int \frac{4x^2}{x^3 + x^2 - x - 1}\,dx$$

To factorize the divisor polynomial, Ruffini is applied

$$
\begin{array}{c|cccc}
 & 1 & 1 & -1 & -1 \\
1 & & 1 & 2 & 1 \\
\hline
 & 1 & 2 & 1 & 0
\end{array}
$$

$$x^3 + x^2 - x - 1 = (x - 1)(x^2 + 2x + 1) = (x - 1)(x + 1)^2$$

Proceed analogously to the previous exercises

$$\frac{4x^2}{x^3 + x^2 - x - 1} = \frac{4x^2}{(x-1)(x+1)^2} = \frac{A}{x-1} + \frac{B}{x+1} + \frac{C}{(x+1)^2}$$

$$4x^2 = A(x+1)^2 + B(x-1)(x+1) + C(x-1)$$

$$4x^2 = A(x^2 + 2x + 1) + B(x-1)(x+1) + C(x-1)$$

$$4x^2 = Ax^2 + 2Ax + A + Bx^2 + Bx - Bx - B + Cx - C$$

$$4x^2 = (A+B)x^2 + (2A+C)x + (A - B - C)$$

Now we proceed to build the system of equations and its resolution

$$
\begin{array}{llll}
A + B & = 4 & \text{Equation 1} & \rightarrow \quad A = 1, \ B = 3 \ y \ C = -2 \\
2A \quad + C & = 0 & \text{Equation 2} \\
A - B - C & = 0 & \text{Equation 3}
\end{array}
$$

$$\int \frac{4x^2}{x^3 + x^2 - x - 1} dx = \int \frac{dx}{x-1} + 3\int \frac{dx}{x+1} - 2\int \frac{dx}{(x+1)^2}$$

$$\int \frac{4x^2}{x^3 + x^2 - x - 1} dx = \ln|x-1| + 3\ln|x+1| + \frac{2}{x+1} + C$$

$$\boxed{\int \frac{4x^2}{x^3 + x^2 - x - 1} dx = \ln|x-1| + 3\ln|x+1| + \frac{2}{x+1} + C}$$

137

$$\int \frac{6x}{x^3 - 8} dx$$

To factor the polynomial of the divisor the formula of the difference of cubes is applied

$$a^3 - b^3 = (a - b)(a^2 + ab + b^2)$$

$$x^3 - 8 = x^3 - 2^3 = (x - 2)(x^2 + 2x + 4)$$

$$\frac{6x}{x^3 - 8} = \frac{A}{x - 2} + \frac{Bx + C}{x^2 + 2x + 4}$$

$6x = A(x^2 + 2x + 4) + (Bx + C)(x - 2)$
$6x = Ax^2 + 2Ax + 4A + Bx^2 - 2Bx + Cx - 2C$
$6x = (A + B)x^2 + (2A - 2B + C)x + (4A - 2C)$

Now we proceed to build the system of equations and its resolution

$$
\begin{array}{llll}
A & + B & & = 0 \qquad \text{Equation 1} \quad \rightarrow \quad A = 1 \,,\, B = -1 \; y \; C = 2 \\
2A & - 2B & + C & = 6 \qquad \text{Equation 2} \\
4A & & - 2C & = 0 \qquad \text{Equation 3}
\end{array}
$$

$$\int \frac{6x}{x^3 - 8} dx = \int \frac{dx}{x - 2} + \int \frac{-x + 2}{x^2 + 2x + 4} dx$$

$$\int \frac{6x}{x^3 - 8} dx = \int \frac{dx}{x - 2} + \int \frac{-x - 1}{x^2 + 2x + 4} dx + \int \frac{3}{x^2 + 2x + 4}$$

$$\int \frac{6x}{x^3 - 8} dx = \int \frac{dx}{x - 2} - \int \frac{x + 1}{x^2 + 2x + 4} dx + 3 \int \frac{dx}{(x + 1)^2 + (\sqrt{3})^2}$$

$$\int \frac{6x}{x^3 - 8} dx = \ln|x - 2| - \frac{1}{2} \ln|x^2 + 2x + 4| + \frac{3}{\sqrt{3}} arctan\frac{(x + 1)}{\sqrt{3}} + C$$

$$\boxed{\int \frac{6x}{x^3 - 8} dx = \ln|x - 2| - \frac{1}{2} \ln|x^2 + 2x + 4| + \sqrt{3} arctan\frac{\sqrt{3}(x + 1)}{3} + C}$$

$$\int \frac{x^2 - x + 9}{(x^2 + 9)^2} dx$$

138

In the divisor of the integrand a quadratic term is presented, therefore, the fraction decomposes like this

$$\frac{x^2 - x + 9}{(x^2 + 9)^2} = \frac{Ax + B}{(x^2 + 9)} + \frac{Cx + D}{(x^2 + 9)^2}$$

$$x^2 - x + 9 = (Ax + B)(x^2 + 9) + Cx + D$$

$$x^2 - x + 9 = Ax^3 + 9Ax + Bx^2 + 9B + Cx + D$$

$$x^2 - x + 9 = Ax^3 + Bx^2 + (9A + C)x + (9B + D)$$

From the previous expression you get

$$
\begin{array}{llll}
9A & + C & = -1 & \text{Equation 1} \\
9B & + D & = 9 & \text{Equation 2}
\end{array}
$$

$A = 0 \; and \; B = 1$

From equation 1 we have C = -1 and from equation 2 D = 0

$$\int \frac{x^2 - x + 9}{(x^2 + 9)^2} dx = \int \frac{dx}{x^2 + 9} - \int \frac{xdx}{(x^2 + 9)^2}$$

$$\int \frac{x^2 - x + 9}{(x^2 + 9)^2} dx = \frac{1}{3} arctan \frac{x}{3} + \frac{1}{2} \frac{1}{x^2 + 9} + C$$

$$\boxed{\int \frac{x^2 - x + 9}{(x^2 + 9)^2} dx = \frac{1}{3} arctan \frac{x}{3} + \frac{1}{2x^2 + 18} + C}$$

139

$$\int \frac{x^2 - 4x + 7}{x^3 - x^2 + x + 3}\, dx$$

To factorize the divisor polynomial of the integrand, Ruffini is applied. Do it!

$$\frac{x^2 - 4x + 7}{x^3 - x^2 + x + 3} = \frac{x^2 - 4x + 7}{(x + 1)(x^2 - 2x + 3)} = \frac{A}{x + 1} + \frac{Bx + C}{x^2 - 2x + 3}$$

$$x^2 - 4x + 7 = A(x^2 - 2x + 3) + (Bx + C)(x + 1)$$

$$x^2 - 4x + 7 = Ax^2 - 2Ax + 3A + Bx^2 + Bx + Cx + C$$

$$x^2 - 4x + 7 = (A + B)x^2 + (-2A + B + C)x + (3A + C)$$

Now we proceed to build the system of equations and its resolution

$$
\begin{array}{|llll}
A & + B & & = 1 \quad \text{Equation 1} \\
-2A & + B & + C & = -4 \quad \text{Equation 2} \\
3A & & + C & = 7 \quad \text{Equation 3} \\
\hline
\end{array}
$$

$\rightarrow \; A = 2,\; B = -1 \; and \; C = 1$

As a result, you get

$$\int \frac{x^2 - 4x + 7}{x^3 - x^2 + x + 3}\, dx = \int \frac{A}{x + 1}\, dx + \int \frac{Bx + C}{x^2 - 2x + 3}\, dx$$

$$\int \frac{x^2 - 4x + 7}{x^3 - x^2 + x + 3}\, dx = 2\int \frac{dx}{x + 1} + \int \frac{-x + 1}{x^2 - 2x + 3}\, dx$$

$$\int \frac{x^2 - 4x + 7}{x^3 - x^2 + x + 3}\, dx = 2\int \frac{dx}{x + 1} - \int \frac{x - 1}{x^2 - 2x + 3}\, dx$$

$$\int \frac{x^2 - 4x + 7}{x^3 - x^2 + x + 3}\, dx = 2\ln|x + 1| - \frac{1}{2}\ln|x^2 - 2x + 3| + C$$

$$\boxed{\int \frac{x^2 - 4x + 7}{x^3 - x^2 + x + 3}\, dx = 2\ln|x + 1| - \frac{1}{2}\ln|x^2 - 2x + 3| + C}$$

$$\int \frac{x^2 + x + 3}{x^4 + 6x^2 + 9}\, dx$$

140

The divisor polynomial of the integrand is factored as a perfect square trinomial.

$$\frac{x^2 + x + 3}{x^4 + 6x^2 + 9} = \frac{x^2 + x + 3}{(x^2 + 3)^2} = \frac{Ax + B}{(x^2 + 3)} + \frac{Cx + D}{(x^2 + 3)^2}$$

$$x^2 + x + 3 = (Ax + B)(x^2 + 3) + Cx + D$$

$$x^2 + x + 3 = Ax^3 + 3Ax + Bx^2 + 3B + Cx + D$$

$$x^2 + x + 3 = Ax^3 + Bx^2 + (3A + C)x + (3B + D)$$

From the previous expression we obtain A = 0 and B = 1. And a system of equations

$$\begin{array}{llll}
3A & + C & = 1 & \textit{Equation 1} \\
& 3B & + D & = 3 & \textit{Equation 2}
\end{array}$$

From equation 1 we have C = 1 and from equation 2 D = 0

$$\int \frac{x^2 + x + 3}{x^4 + 6x^2 + 9}\, dx = \int \frac{Ax + B}{x^2 + 3}\, dx + \int \frac{Cx + D}{(x^2 + 3)^2}\, dx$$

$$\int \frac{x^2 + x + 3}{x^4 + 6x^2 + 9}\, dx = \int \frac{dx}{x^2 + 3} + \int \frac{x}{(x^2 + 3)^2}\, dx$$

$$\int \frac{x^2 + x + 3}{x^4 + 6x^2 + 9}\, dx = \int \frac{dx}{x^2 + (\sqrt{3})^2} + \int \frac{x}{(x^2 + 3)^2}\, dx$$

$$\int \frac{x^2 + x + 3}{x^4 + 6x^2 + 9}\, dx = \frac{1}{\sqrt{3}} arctan \frac{x}{\sqrt{3}} - \frac{1}{2}\frac{1}{x^2 + 3} + C$$

$$\boxed{\int \frac{x^2 + x + 3}{x^4 + 6x^2 + 9}\, dx = \frac{\sqrt{3}}{3} arctan \frac{x\sqrt{3}}{3} - \frac{1}{2(x^2 + 3)} + C}$$

141

$$\int \frac{4x}{(x^2 + 1)(x^2 + 2x + 3)}\, dx$$

Now we proceed to decompose the integrand into partial fractions, since the denominator presents the product of two quadratic factors

$$\frac{4x}{(x^2 + 1)(x^2 + 2x + 3)} = \frac{Ax + B}{(x^2 + 1)} + \frac{Cx + D}{(x^2 + 2x + 3)}$$

$$\frac{4x}{(x^2 + 1)(x^2 + 2x + 3)} = \frac{(Ax + B)(x^2 + 2x + 3) + (Cx + D)(x^2 + 1)}{(x^2 + 1)(x^2 + 2x + 3)}$$

From the previous expression you have

$$4x = (Ax + B)(x^2 + 2x + 3) + (Cx + D)(x^2 + 1)$$

Removing parentheses

$$4x = Ax^3 + 2Ax^2 + 3Ax + Bx^2 + 2Bx + 3B + Cx^3 + Cx + Dx^2 + D$$

Terms are grouped with their respective power of x.

$$0x^3 + 0x^2 + 4x + 0 = (A + C)x^3 + (2A + B + D)x^2 + (3A + 2B + C)x + (3B + D)$$

Now the coefficients of the powers of x are compared and the system of equations is constructed:

A	$+ C$		$= 0$	*Equation 1*
$2A + B$		$+ D$	$= 0$	*Equation 2*
$3A + 2B + C$			$= 4$	*Equation 3*
$3B$		$+ D$	$= 0$	*Equation 4*

Using a programmable scientific calculator you get the following values

$$A = 1 \quad , \quad B = 1 \quad , \quad C = -1 \quad and \quad D = -3$$

Consequently, you have

$$\int \frac{4x}{(x^2 + 1)(x^2 + 2x + 3)}\, dx = \int \frac{Ax + B}{(x^2 + 1)}\, dx + \int \frac{Cx + D}{(x^2 + 2x + 3)}\, dx$$

$$\int \frac{4x}{(x^2+1)(x^2+2x+3)}\,dx = \int \frac{x+1}{x^2+1}\,dx - \int \frac{x+3}{x^2+2x+3}\,dx \quad \textbf{\textit{Ec.}}\,\textbf{1}$$

The integrals are solved separately from equation 1.

$$\int \frac{x+1}{x^2+1}\,dx$$

To work with u and du it is necessary to separate the integral into two integrals

$$\int \frac{x+1}{x^2+1}\,dx = \int \frac{x\,dx}{x^2+1} + \int \frac{dx}{x^2+1}$$

For the integral $\int \frac{x\,dx}{x^2+1}$: $u = x^2+1 \longrightarrow du = 2x\,dx$. And the next integral its result is an inverse tangent.

$$\int \frac{x+1}{x^2+1}\,dx = \frac{1}{2}\ln|x^2+1| + \arctan x + C_1$$

$$\int \frac{x+1}{x^2+1}\,dx = \frac{1}{2}\ln|x^2+1| + \arctan x + C_1 \quad \textbf{\textit{Eq.}}\,\textbf{2}$$

$$\int \frac{x+3}{x^2+2x+3}\,dx$$

The square is completed and the integral is separated into two integrals

$$x^2+2x+3 = x^2+2x+1+3-1$$

$$x^2+2x+3 = (x+1)^2+2$$

$$\int \frac{x+3}{x^2+2x+3}\,dx = \int \frac{x+1+2}{(x+1)^2+2} = \int \frac{x+1}{(x+1)^2+2}\,dx + 2\int \frac{dx}{(x+1)^2+(\sqrt{2})^2}\,dx$$

For the first integral $\int \frac{x+1}{(x+1)^2+2}\,dx$: $u = (x+1)^2+2$, $du = 2(x+1)dx$. And the second integral $\int \frac{dx}{(x+1)^2+(\sqrt{2})^2}\,dx$: its result is a reverse tangent.

$$\int \frac{x+3}{x^2+2x+3}\,dx = \frac{1}{2}\ln|(x+1)^2+2| + \frac{1}{\sqrt{2}}\arctan\frac{x+1}{\sqrt{2}} + C_2$$

$$\int \frac{x+3}{x^2+2x+3}\,dx = \frac{1}{2}\ln|(x+1)^2+2| + \frac{2}{\sqrt{2}}\arctan\frac{x+1}{\sqrt{2}} + C_2 \;\; \textbf{Eq. 3}$$

Substituting equation 2 and equation 3 in equation 1 and taking into account that $C = C_1 + C_2$ is obtained

$$\int \frac{4x}{(x^2+1)(x^2+2x+3)}\,dx = \int \frac{x+1}{x^2+1}\,dx - \int \frac{x+3}{x^2+2x+3}\,dx$$

$$\int \frac{4x}{(x^2+1)(x^2+2x+3)}\,dx = \frac{1}{2}\ln|x^2+1| + \arctan x + C_1 - \frac{1}{2}\ln|(x+1)^2+2| - \frac{2}{\sqrt{2}}\arctan\frac{x+1}{\sqrt{2}} + C_2$$

$$\int \frac{4x}{(x^2+1)(x^2+2x+3)}\,dx = \frac{1}{2}\ln\left(\frac{x^2+1}{x^2+2x+3}\right) + \arctan x - \sqrt{2}\arctan\frac{x+1}{\sqrt{2}} + C$$

$$\boxed{\int \frac{4x}{(x^2+1)(x^2+2x+3)}\,dx = \frac{1}{2}\ln\left(\frac{x^2+1}{x^2+2x+3}\right) + \arctan x - \sqrt{2}\arctan\frac{x+1}{\sqrt{2}} + C}$$

142

$$\int \frac{x^2}{(x^2+4)^2}\,dx$$

The integrand is broken down into partial fractions and is operated mathematically.

$$\frac{x^2}{(x^2+4)^2} = \frac{Ax+B}{(x^2+4)} + \frac{Cx+D}{(x^2+4)^2}$$

$$x^2 = (Ax+B)(x^2+4) + Cx + D$$
$$x^2 = Ax^3 + 4Ax + Bx^2 + 4B + Cx + D$$
$$0x^3 + x^2 + 0x + 0 = Ax^3 + Bx^2 + (4A+C)x + (4B+D)$$

Now the coefficients of the powers of x are compared and the system of equations is constructed

$$
\begin{aligned}
A & = 0 & \text{Equation 1}\\
B & = 1 & \text{Equation 2}\\
4A + C & = 0 & \text{Equation 3}\\
4B + D & = 0 & \text{Equation 4}
\end{aligned}
$$

Using a programmable scientific calculator you get the following values

$$A = 0, \quad B = 1, \quad C = 0 \quad and \quad D = -4$$

Therefore,

$$\int \frac{x^2}{(x^2 + 4)^2}\, dx = \int \frac{dx}{x^2 + 4} - 4\int \frac{dx}{(x^2 + 4)^2} \qquad \textbf{Equation 1}$$

The integral $\int \frac{dx}{x^2+4}$ es una tangente inversa. the siguient integral $\int \frac{dx}{(x^2+4)^2}$ it is solved by trigonometric substitution.

$$\int \frac{dx}{x^2 + 4} = \frac{1}{2}\arctan\frac{x}{2} + C_1 \quad \textit{Equation 2}$$

$$\int \frac{dx}{(x^2 + 4)^2} \qquad \textit{The integral is rewritten as follows}$$

$$\int \frac{dx}{(x^2 + 4)^2} = \int \frac{dx}{(\sqrt{x^2 + 2^2})^4} \qquad \begin{aligned} &\textit{Where:}\\ &x = 2\tan\theta\,, dx = 2\sec^2\theta d\theta\,, \sqrt{x^2 + 2^2} = 2\sec\theta \end{aligned}$$

In the topic that corresponds to the resolution of integrals by trigonometric substitution, the reader will find reference exercises to solve the integral one.

Therefore, the result of the integral $\int \frac{dx}{(x^2+4)^2}$ is

$$\int \frac{dx}{(x^2 + 4)^2} = \frac{1}{16}\left(\arctan\frac{x}{2} + \frac{2x}{x^2 + 4}\right) + C_2 \quad \textbf{Equation 3}$$

Substituting equation 2 and equation 3 in equation 1 and taking into account that $C = C_1 + C_2$, is obtained

$$\int \frac{x^2}{(x^2+4)^2}\, dx = \int \frac{dx}{x^2+4} - 4 \int \frac{dx}{(x^2+4)^2}$$

$$\int \frac{x^2}{(x^2+4)^2}\, dx = \frac{1}{2} \arctan \frac{x}{2} - 4 \left(\frac{1}{16} \left(\arctan \frac{x}{2} + \frac{2x}{x^2+4} \right) \right) + C$$

$$\int \frac{x^2}{(x^2+4)^2}\, dx = \frac{1}{2} \arctan \frac{x}{2} - \frac{1}{4} \arctan \frac{x}{2} - \frac{1}{2}\, \frac{x}{x^2+4} + C$$

$$\boxed{\int \frac{x^2}{(x^2+4)^2}\, dx = \frac{1}{4} \arctan \frac{x}{2} - \frac{x}{2(x^2+4)} + C}$$

143 $$\boxed{\int \frac{sen\,\theta\, d\theta}{cos^2 + cos\theta - 2}}$$

The denominator of the fraction is factored

$$cos^2 + cos\theta - 2 = (cos\theta + 2)(cos\theta - 1)$$

$$\int \frac{sen\,\theta\, d\theta}{cos^2 + cos\theta - 2} = \int \frac{sen\,\theta\, d\theta}{(cos\theta + 2)(cos\theta - 1)}$$

We work both $cos\theta + 2$ and $cos\theta - 1$ as a function of u, that is,

$$u = cos\theta + 2 \longrightarrow du = -sen\theta\, d\theta$$

Yes $cos\theta = u - 2$ so $cos\theta - 1 = u - 2 - 1 \rightarrow cos\theta - 1 = u - 3$

$$\int \frac{sen\,\theta\, d\theta}{cos^2 + cos\theta - 2} = -\int \frac{du}{u(u-3)}$$

Now the integrand is broken down into partial fractions and mathematically operated:

$$\frac{1}{u(u-3)} = \frac{A}{u} + \frac{B}{(u-3)}$$

$$1 = A(u-3) + B(u)$$

$$1 = Au - 3A + Bu$$

$$1 = (A + B)u - 3A$$

$$-3A = 1 \rightarrow A = -\frac{1}{3}$$

$$A + B = 0 \quad \rightarrow \quad B = \frac{1}{3}$$

Consequently, you have

$$-\int \frac{du}{u(u-3)} = -\left(-\frac{1}{3}\int \frac{du}{u} + \frac{1}{3}\int \frac{du}{u-3}\right)$$

$$\int \frac{sen\,\theta\,d\theta}{cos^2 + cos\theta - 2} = \frac{1}{3}ln|cos\theta + 2| - \frac{1}{3}ln|cos\theta + 2 - 3| + C$$

$$\int \frac{sen\,\theta\,d\theta}{cos^2 + cos\theta - 2} = \frac{1}{3}ln|cos\theta + 2| - \frac{1}{3}ln|cos\theta - 1| + C$$

Because the $|cos\theta| < 1$ has

$$|cos\theta - 1| = |1 - cos\theta|$$

Therefore, you get

$$\int \frac{sen\,\theta\,d\theta}{cos^2 + cos\theta - 2} = \frac{1}{3}ln|2 + cos\theta| - \frac{1}{3}ln|1 - cos\theta| + C$$

$$\int \frac{sen\,\theta\,d\theta}{cos^2 + cos\theta - 2} = \frac{1}{3}ln\left|\frac{2 + cos\theta}{1 - cos\theta}\right| + C$$

$$\boxed{\int \frac{\boldsymbol{sen\,\theta\,d\theta}}{\boldsymbol{cos^2 + cos\theta - 2}} = \frac{1}{3}\boldsymbol{ln}\left|\frac{\boldsymbol{2 + cos\theta}}{\boldsymbol{1 - cos\theta}}\right| + \boldsymbol{C}}$$

144

$$\int \frac{e^x dx}{e^{2x} + 3e^x + 2}$$

The denominator of the fraction is rewritten and then factored

$$e^{2x} + 3e^x + 2 = (e^x)^2 + 3e^x + 2 = (e^x + 2)(e^x + 1)$$

$$\int \frac{e^x dx}{e^{2x} + 3e^x + 2} = \int \frac{e^x dx}{(e^x + 2)(e^x + 1)}$$

It operates analogously to the previous case.

$$\int \frac{e^x dx}{(e^x + 2)(e^x + 1)} = \int \frac{du}{u(u - 1)}$$

Now the integrand is broken down into partial fractions and mathematically operated

$$\frac{1}{u(u - 1)} = \frac{A}{u} + \frac{B}{u - 1}$$

$$1 = A(u - 1) + B(u)$$

$$1 = Au - A + Bu$$

$$1 = (A + B)u - A$$

$$-A = 1 \rightarrow \boldsymbol{A = -1} \quad and \quad A + B = 0 \rightarrow \boldsymbol{B = 1}$$

$$\int \frac{du}{u(u - 1)} = -\int \frac{du}{u} + \int \frac{du}{u - 1}$$

$$\int \frac{e^x dx}{e^{2x} + 3e^x + 2} = -ln|e^x + 2| + ln|e^x + 1| + C$$

$$\boxed{\int \frac{e^x dx}{e^{2x} + 3e^x + 2} = ln \left| \frac{e^x + 1}{e^x + 2} \right| + C}$$

$$\int \frac{(2 + \tan^2\theta)\sec^2\theta \; d\theta}{1 + \tan^3\theta}$$

145

The initial question is the following: Can this Integral be solved by applying integration of rational functions? The answer is yes. So, what is the procedure? The first thing is to transform the trigonometric terms of the integrand in terms of u and du. Then the denominator is factored to decompose it into partial fractions.

Do

$$u = \tan\theta \; \longrightarrow \; du = \sec^2\theta \; d\theta$$

It is replaced in the integral:

$$\int \frac{(2 + \tan^2\theta)\sec^2\theta \; d\theta}{1 + \tan^3\theta} = \int \frac{(2 + u^2) \; du}{1 + u^3}$$

Then the denominator is factored:

$$1 + u^3 = (1 + u)(1 - u + u^2) = (1 + u)(u^2 - u + 1)$$

$$\int \frac{(2 + u^2) \; du}{1 + u^3} = \int \frac{(2 + u^2) \; du}{(1 + u)(u^2 - u + 1)}$$

Now the integrand is broken down into simple fractions and mathematically operated:

$$\frac{(2 + u^2)}{(1 + u)(u^2 - u + 1)} = \frac{A}{1 + u} + \frac{Bu + C}{u^2 - u + 1}$$

$$2 + u^2 = A(u^2 - u + 1) + (Bu + C)(1 + u)$$

$$2 + u^2 = Au^2 - Au + A + Bu + Bu^2 + C + Cu$$

$$u^2 + 0u + 2 = (A + B)u^2 + (-A + B + C)u + (A + C)$$

$$
\begin{array}{lllll}
A & + & B & & = 1 & \textit{Equation 1} \\
-A & + & B & + C & = 0 & \textit{Equation 2} \\
A & & & + C & = 2 & \textit{Equation 3} \\
\hline
\end{array}
$$

Using a programmable scientific calculator you get the following values

$$A = 1, \quad B = 0 \quad and \quad C = 1$$

Therefore,

$$\int \frac{(2 + u^2)\, du}{(1 + u)(u^2 - u + 1)} = \int \frac{du}{1 + u} + \int \frac{du}{u^2 - u + 1}$$

The square is completed in the second integral. In consecuense

$$\int \frac{(2 + u^2)\, du}{(1 + u)(u^2 - u + 1)} = ln|1 + u| + \frac{1}{\frac{\sqrt{3}}{2}} arctan \frac{u - \frac{1}{2}}{\frac{\sqrt{3}}{2}} + C$$

$$\int \frac{(2 + tan^2\theta)sec^2\theta\, d\theta}{1 + tan^3\theta} = ln|1 + tan\theta| + \frac{1}{\frac{\sqrt{3}}{2}} arctan \frac{tan\theta - \frac{1}{2}}{\frac{\sqrt{3}}{2}} + C$$

$$\int \frac{(2 + tan^2\theta)sec^2\theta\, d\theta}{1 + tan^3\theta} = ln|1 + tan\theta| + \frac{2\sqrt{3}}{3} arctan \frac{\frac{2tan\theta - 1}{2}}{\frac{\sqrt{3}}{2}} + C$$

$$\int \frac{(2 + tan^2\theta)sec^2\theta\, d\theta}{1 + tan^3\theta} = ln|1 + tan\theta| + \frac{2\sqrt{3}}{3} arctan \frac{2tan\theta - 1}{\sqrt{3}} + C$$

$$\boxed{\int \frac{(2 + tan^2\theta)sec^2\theta\, d\theta}{1 + tan^3\theta} = ln|1 + tan\theta| + \frac{2\sqrt{3}}{3} arctan \frac{2tan\theta - 1}{\sqrt{3}} + C}$$

$$\int \frac{3x^4\,dx}{(x^2+1)^2}$$

146

Observe that when making the square in the divisor, x^4 is obtained in the polynomial, equal to the numerator. The division of the polynomials is then carried out. With this procedure, the integral problem is simplified.

$$(x^2+1)^2 = x^4 + 2x^2 + 1$$

$$\frac{3x^4}{x^4+2x^2+1} = 3 - \frac{6x^2+3}{x^4+2x^2+1}$$

Consequently, you have

$$\int \frac{3x^4\,dx}{(x^2+1)^2} = 3\int dx - 3\int \frac{2x^2+1}{x^4+2x^2+1}\,dx$$

$$\int \frac{3x^4\,dx}{(x^2+1)^2} = 3x - 3\int \frac{2x^2+1}{(x^2+1)^2}\,dx \quad \textbf{\textit{Equation 1}}$$

Now we proceed to solve the integral of equation 1

$$\int \frac{2x^2+1}{(x^2+1)^2}\,dx = \int \frac{Ax+B}{x^2+1}\,dx + \int \frac{Cx+D}{(x^2+1)^2}\,dx$$

Then proceed to calculate the respective values of: A, B, C and D.

$$\frac{2x^2+1}{(x^2+1)^2} = \frac{Ax+B}{x^2+1} + \frac{Cx+D}{(x^2+1)^2}$$

$$2x^2+1 = (Ax+B)(x^2+1) + Cx + D$$

$$2x^2+1 = Ax^3 + Ax + Bx^2 + B + Cx + D$$

$$2x^2+1 = Ax^3 + Bx^2 + (A+C)x + (B+D)$$

Next, the coefficients of the powers of x are compared.

$$A = 0, B = 2, A + C = 0 \rightarrow C = 0, \ B + D = 1 \rightarrow D = -1$$

$$\int \frac{2x^2 + 1}{(x^2 + 1)^2} dx = 2 \int \frac{dx}{x^2 + 1} - \int \frac{dx}{(x^2 + 1)^2}$$

The first integral integral is immediate. The other integral is solved by trigonometric substitution. For this resolution, take as a reference the problem solved 142.

$$\int \frac{2x^2 + 1}{(x^2 + 1)^2} dx = 2 \arctan x - \frac{1}{2} (\arctan x + \frac{x}{x^2 + 1}) + C_1$$

$$\int \frac{2x^2 + 1}{(x^2 + 1)^2} dx = 2 \arctan x - \frac{1}{2} \arctan x - \frac{x}{2(x^2 + 1)} + C_1$$

$$\int \frac{2x^2 + 1}{(x^2 + 1)^2} dx = \frac{3}{2} \arctan x - \frac{x}{2(x^2 + 1)} + C_1$$

Substituting the previous result in equation 1 we have:

$$\int \frac{3x^4 \, dx}{(x^2 + 1)^2} = 3x - 3 \int \frac{2x^2 + 1}{(x^2 + 1)^2} dx$$

$$\int \frac{3x^4 \, dx}{(x^2 + 1)^2} = 3x - \frac{9}{2} \arctan x + \frac{3x}{2(x^2 + 1)} + C$$

$$\boxed{\int \frac{3x^4 \, dx}{(x^2 + 1)^2} = 3x - \frac{9}{2} \arctan x + \frac{3x}{2(x^2 + 1)} + C}$$

$$\int \frac{(2x^2 + 3x - 1)\, dx}{x^3 + 3x^2 + 4x + 2}$$

147

The divisor of the fraction is factored.

	1	3	4	2
-1		-1	-2	-2
	1	2	2	0

$$x^3 + 3x^2 + 4x + 2 = (x + 1)(x^2 + 2x + 2)$$

Consequently, you have

$$\int \frac{(2x^2 + 3x - 1)\, dx}{x^3 + 3x^2 + 4x + 2)} = \int \frac{(2x^2 + 3x - 1)\, dx}{(x + 1)(x^2 + 2x + 2)} = \int \frac{A}{x + 1} + \int \frac{Bx + C}{x^2 + 2x + 2}$$

We proceed to calculate the values of A, B and C.

$$\frac{(2x^2 + 3x - 1)}{(x + 1)(x^2 + 2x + 2)} = \frac{A}{x + 1} + \frac{Bx + C}{x^2 + 2x + 2}$$

$$2x^2 + 3x - 1 = A(x^2 + 2x + 2) + (Bx + C)(x + 1)$$

$$2x^2 + 3x - 1 = Ax^2 + 2xA + 2A + Bx^2 + Bx + Cx + C$$

$$2x^2 + 3x - 1 = (A + B)x^2 + (2A + B + C)x + (2A + C)$$

$$\begin{vmatrix} A & + B & & = & 2 \\ 2A & + B & + C & = & 3 \\ 2A & & + C & = & -1 \end{vmatrix}$$

Using a programmable scientific calculator you get the following values

$$A = -2 \ , \quad B = 4 \quad y \quad C = 3$$

$$\int \frac{(2x^2 + 3x - 1)\, dx}{x^3 + 3x^2 + 4x + 2)} = -2 \int \frac{dx}{x + 1} + \int \frac{(4x + 3)dx}{x^2 + 2x + 2}$$

$$\int \frac{(2x^2 + 3x - 1)\,dx}{x^3 + 3x^2 + 4x + 2)} = -2\int \frac{dx}{x + 1} + 2\int \frac{(2x + 2) - 1}{x^2 + 2x + 2}\,dx$$

$$\int \frac{(2x^2 + 3x - 1)\,dx}{x^3 + 3x^2 + 4x + 2)} = -2\int \frac{dx}{x + 1} + 2\int \frac{(2x + 2)\,dx}{x^2 + 2x + 2} - 2\int \frac{dx}{x^2 + 2x + 2}$$

$$\int \frac{(2x^2 + 3x - 1)\,dx}{x^3 + 3x^2 + 4x + 2)} = -2ln|x + 1| + 2ln|x^2 + 2x + 2| - 2\arctan(x + 1) + C$$

$$\boxed{\int \frac{(2x^2 + 3x - 1)\,dx}{x^3 + 3x^2 + 4x + 2)} = -2ln|x + 1| + 2ln|x^2 + 2x + 2| - 2\arctan(x + 1) + C}$$

148

$$\int \frac{(3x^2 + x - 2)\,dx}{(x - 1)(x^2 + 1)}$$

Now we proceed to decompose the integrand into partial fractions, since the denominator presents the product of two factors, one linear and the other quadratic.

$$\frac{(3x^2 + x - 2)}{(x - 1)(x^2 + 1)} = \frac{A}{x - 1} + \frac{Bx + C}{x^2 + 1}$$

$$3x^2 + x - 2 = A(x^2 + 1) + (Bx + C)(x - 1)$$

$$3x^2 + x - 2 = Ax^2 + A + Bx^2 - Bx + Cx - C$$

$$3x^2 + x - 2 = (A + B)x^2 + (-B + C)x + (A - C)$$

$$\begin{vmatrix} A + B & = 3 \\ -B + C & = 1 \\ A & -C = -2 \end{vmatrix} \quad \rightarrow \quad \boldsymbol{A = 1,\ B = 2\ and\ C = 3}$$

$$\int \frac{(3x^2 + x - 2)\,dx}{(x - 1)(x^2 + 1)} = \int \frac{A}{x - 1} + \int \frac{Bx + C}{x^2 + 1}$$

$$\int \frac{(3x^2 + x - 2)dx}{(x - 1)(x^2 + 1)} = \int \frac{dx}{x - 1} + \int \frac{(2x + 3)dx}{x^2 + 1}$$

$$\int \frac{(3x^2 + x - 2)dx}{(x - 1)(x^2 + 1)} = \int \frac{dx}{x - 1} + \int \frac{2x\,dx}{x^2 + 1} + 3\int \frac{dx}{x^2 + 1}$$

$$\boxed{\int \frac{(3x^2 + x - 2)dx}{(x - 1)(x^2 + 1)} = ln\left|(x - 1)(x^2 + 1)\right| + 3\arctan x + C}$$

$$\boxed{\int \frac{2x^2 - 7x - 1}{x^3 + x^2 - x - 1}dx}$$

149

The divisor of the fraction is factored.

	1	1	-1	-1
		1	2	1
1				
	1	2	1	0

$$x^3 + x^2 - x - 1 = (x - 1)(x^2 + 2x + 1) = (x - 1)(x + 1)^2$$

Consequently, you have

$$\frac{2x^2 - 7x - 1}{(x - 1)(x + 1)^2} = \frac{A}{x - 1} + \frac{B}{x + 1} + \frac{C}{(x + 1)^2}$$

$$2x^2 - 7x - 1 = A(x + 1)^2 + B(x - 1)(x + 1) + C(x - 1)$$

$$2x^2 - 7x - 1 = A(x^2 + 2x + 1) + Bx^2 + Bx - Bx - B + Cx - C$$

$$2x^2 - 7x - 1 = Ax^2 + 2Ax + A + Bx^2 - B + Cx - C$$

$$2x^2 - 7x - 1 = (A + B)x^2 + (2A + C)x + (A - B - C)$$

Next, the coefficients of the powers of x are compared and the system of equations is constructed.

$$\begin{array}{rrrrcr} A & + B & & = & 2 \\ 2A & & + C & = & -7 \\ A & - B & - C & = & -1 \end{array}$$

Using a programmable scientific calculator you get the following values

$$A = -\frac{3}{2}, \quad B = \frac{7}{2} \quad and \quad C = -4$$

$$\int \frac{2x^2 - 7x - 1}{x^3 + x^2 - x - 1} dx = \int \frac{A}{x - 1} + \int \frac{B}{x + 1} + \int \frac{C}{(x + 1)^2}$$

$$\int \frac{2x^2 - 7x - 1}{x^3 + x^2 - x - 1} dx = -\frac{3}{2} \int \frac{dx}{x - 1} + \frac{7}{2} \int \frac{dx}{x + 1} - 4 \int \frac{dx}{(x + 1)^2}$$

$$\int \frac{2x^2 - 7x - 1}{x^3 + x^2 - x - 1} dx = -\frac{3}{2} \ln|x - 1| + \frac{7}{2} \ln|x + 1| + \frac{4}{x + 1} + C$$

$$\int \frac{2x^2 - 7x - 1}{x^3 + x^2 - x - 1} dx = \frac{1}{2} [-3\ln|x - 1| + 7\ln|x + 1|] + \frac{4}{x + 1} + C$$

$$\boxed{\int \frac{2x^2 - 7x - 1}{x^3 + x^2 - x - 1} dx = \frac{1}{2} \ln \left| \frac{(x + 1)^7}{(x - 1)^3} \right| + \frac{4}{x + 1} + C}$$

$$\int \frac{dx}{e^{2x} + e^x - 2}$$

150

The divisor is rewritten and squares are completed

$$e^{2x} + e^x - 2 = (e^x)^2 + e^x - 2 = (e^x)^2 + e^x + \frac{1}{4} - 2 - \frac{1}{4}$$

$$e^{2x} + e^x - 2 = \left[(e^x)^2 + e^x + \frac{1}{4}\right] - \frac{9}{4} = \left(e^x + \frac{1}{2}\right)^2 - \left(\frac{3}{2}\right)^2$$

$$\int \frac{dx}{e^{2x} + e^x - 2} = \int \frac{dx}{\left(e^x + \frac{1}{2}\right)^2 - \left(\frac{3}{2}\right)^2}$$

Now we proceed to transform the exponential term as a function of u and du.

Be $\qquad u = e^x + \frac{1}{2} \;\to\; du = e^x dx$

It clears $\qquad dx = \frac{du}{e^x}$

As $\qquad e^x = u - \frac{1}{2}$

You have $\qquad dx = \frac{du}{u - \frac{1}{2}}$

Then substitutions are made

$$\int \frac{dx}{\left(e^x + \frac{1}{2}\right)^2 - \left(\frac{3}{2}\right)^2} = \int \frac{\frac{du}{u - \frac{1}{2}}}{(u)^2 - \left(\frac{3}{2}\right)^2} = \int \frac{du}{\left(u - \frac{1}{2}\right)\left(u + \frac{3}{2}\right)\left(u - \frac{3}{2}\right)}$$

Now the integrand is broken down into simple fractions and mathematically operated

$$\frac{1}{\left(u - \frac{1}{2}\right)\left(u + \frac{3}{2}\right)\left(u - \frac{3}{2}\right)} = \frac{A}{\left(u - \frac{1}{2}\right)} + \frac{B}{\left(u + \frac{3}{2}\right)} + \frac{C}{\left(u - \frac{3}{2}\right)}$$

$$1 = A\left(u + \frac{3}{2}\right)\left(u - \frac{3}{2}\right) + B\left(u - \frac{1}{2}\right)\left(u - \frac{3}{2}\right) + C\left(u - \frac{1}{2}\right)\left(u + \frac{3}{2}\right)$$

Comment: Another procedure to calculate the values of A, B and C is to substitute values suitable for (u) in the fundamental equation to obtain zeros in the other factors.

For $u = \frac{1}{2}$ it is determined that the factors that cancel are B and C. Therefore:

$1 = A(2)(-1)$. Which implies that $A = -\frac{1}{2}$

For $u = -\frac{3}{2}$ it is determined that the factors that are annulled are the A and C. Therefore:

$1 = B(-2)(-3)$. Which implies that $B = \frac{1}{6}$

For $u = \frac{3}{2}$ it is determined that the factors that are annulled are A and B. Therefore:

$1 = C(1)(3)$. Which implies that $C = \frac{1}{3}$

Consequently, you have

$$\int \frac{du}{\left(u - \frac{1}{2}\right)\left(u + \frac{3}{2}\right)\left(u - \frac{3}{2}\right)} = \int \frac{A\,du}{\left(u - \frac{1}{2}\right)} + \int \frac{B\,du}{\left(u + \frac{3}{2}\right)} + \int \frac{C\,du}{\left(u - \frac{3}{2}\right)}$$

$$\int \frac{du}{\left(u - \frac{1}{2}\right)\left(u + \frac{3}{2}\right)\left(u - \frac{3}{2}\right)} = -\frac{1}{2}\int \frac{du}{\left(u - \frac{1}{2}\right)} + \frac{1}{6}\int \frac{du}{\left(u + \frac{3}{2}\right)} + \frac{1}{3}\int \frac{du}{\left(u - \frac{3}{2}\right)}$$

$$= -\frac{1}{2}\ln\left|u - \frac{1}{2}\right| + \frac{1}{6}\ln\left|u + \frac{3}{2}\right| + \frac{1}{3}\ln\left|u - \frac{3}{2}\right| + C$$

$$= \frac{1}{6}\left[-3\ln\left|u - \frac{1}{2}\right| + \ln\left|u + \frac{3}{2}\right| + 2\ln\left|u - \frac{3}{2}\right|\right] + C$$

$$= \frac{1}{6}\ln\left|\frac{\left(u + \frac{3}{2}\right)\left(u - \frac{3}{2}\right)^2}{\left(u - \frac{1}{2}\right)^3}\right| + C$$

As $u = e^x + \frac{1}{2}$ and operating mathematically you have

$$\int \frac{dx}{e^{2x} + e^x - 2} = \frac{1}{6}\ln\left|\frac{(e^x + 2)(e^x - 1)^2}{e^{3x}}\right| + C$$

$$\boxed{\int \frac{dx}{e^{2x} + e^x - 2} = \frac{1}{6}\ln\left|\frac{(e^x + 2)(e^x - 1)^2}{e^{3x}}\right| + C}$$

In the following exercises are the integrals of rational functions of sine and cosine

$$\int \frac{dx}{1 + sen\, x + \cos x}$$

151

For the resolution of integrals containing sine and cosine functions, the following substitutions will be made.

$$sen\, x = \frac{2u}{1 + u^2} \quad , \quad \cos x = \frac{1 - u^2}{1 + u^2} \quad , \quad dx = \frac{2}{1 + u^2} du \quad , \quad u = tan\frac{x}{2}$$

$$\int \frac{dx}{1 + sen\, x + \cos x} = \int \frac{\dfrac{2}{1 + u^2}}{\dfrac{1 + u^2 + 2u + 1 - u^2}{1 + u^2}} du$$

$$= \int \frac{\dfrac{2}{1 + u^2}}{\dfrac{2u + 2}{1 + u^2}} du = \int \frac{2(1 + u^2)}{(1 + u^2)(2u + 2)}$$

$$= \int \frac{2}{(2u + 2)} du = \int \frac{2}{2(u + 1)} du = \int \frac{du}{u + 1}$$

$$\int \frac{dx}{1 + sen\, x + \cos x} = ln|u + 1| + C$$

As $u = tan\dfrac{x}{2}$ The result of the integral problem is:

$$\boxed{\int \frac{dx}{1 + sen\, x + \cos x} = ln\left|tan\frac{x}{2} + 1\right| + C}$$

152

$$\int \frac{dx}{2 + \cos x}$$

For the resolution of integrals containing sine and cosine functions, the following substitutions will be made.

$$sen\, x = \frac{2u}{1 + u^2} \quad , \quad \cos x = \frac{1 - u^2}{1 + u^2} \quad , \quad dx = \frac{2}{1 + u^2} du \quad , \quad u = \tan\frac{x}{2}$$

$$\int \frac{dx}{2 + \cos x} = \int \frac{\dfrac{2}{1 + u^2}}{2 + \dfrac{1 - u^2}{1 + u^2}} du = \int \frac{\dfrac{2}{1 + u^2}}{\dfrac{2(1 + u^2) + (1 - u^2)}{1 + u^2}} du$$

$$\int \frac{dx}{2 + \cos x} = \int \frac{\dfrac{2}{1 + u^2}}{\dfrac{2 + 2u^2 + 1 - u^2}{1 + u^2}} du = \int \frac{2}{u^2 + 3} du$$

The integrando is rewritten as follows: $\quad u^2 + 3 \;\rightarrow\; u^2 + (\sqrt{3})^2$

$$\int \frac{dx}{2 + \cos x} = \int \frac{2}{u^2 + 3} du = \int \frac{2}{u^2 + (\sqrt{3})^2} du = \frac{2}{\sqrt{3}} arctan\frac{u}{\sqrt{3}} + C$$

As $u = \tan\dfrac{x}{2}$ the result of the integral problem is

$$\int \frac{dx}{2 + \cos x} = \frac{2\sqrt{3}}{3} arctan\frac{\tan\dfrac{x}{2}}{\sqrt{3}} + C$$

$$\boxed{\int \frac{dx}{2 + \cos x} = \frac{2\sqrt{3}}{3} arctan\left[\frac{\sqrt{3}}{3}\tan\frac{x}{2}\right] + C}$$

$$\int \frac{dx}{1 + \sec x}$$

The integral is rewritten

$$\int \frac{dx}{1 + \sec x} = \int \frac{dx}{1 + \dfrac{1}{\cos x}}$$

For the resolution of integrals containing sine and cosine functions, the following substitutions will be made.

$$sen\, x = \frac{2u}{1 + u^2} \quad , \quad \cos x = \frac{1 - u^2}{1 + u^2} \quad , \quad dx = \frac{2}{1 + u^2} du \quad , \quad u = tan \frac{x}{2}$$

$$\int \frac{dx}{1 + \dfrac{1}{\cos x}} = \int \frac{\dfrac{2}{1 + u^2} du}{1 + \dfrac{1}{\dfrac{1 - u^2}{1 + u^2}}} = \int \frac{\dfrac{2}{1 + u^2} du}{1 + \dfrac{1 + u^2}{1 - u^2}} = \int \frac{\dfrac{2}{1 + u^2} du}{\dfrac{1 - u^2 + 1 + u^2}{1 - u^2}}$$

$$\int \frac{dx}{1 + \dfrac{1}{\cos x}} = \int \frac{\dfrac{2}{1 + u^2} du}{\dfrac{2}{1 - u^2}} \int \frac{2(1 - u^2)du}{2(1 + u^2)} = \int \frac{(1 - u^2)du}{1 + u^2}$$

$$\int \frac{dx}{1 + \dfrac{1}{\cos x}} = -\int \frac{(u^2 - 1)du}{u^2 + 1} = -\int \left(1 - \frac{2}{1 + u^2}\right) du$$

$$\int \frac{dx}{1 + \dfrac{1}{\cos x}} = -\int du + 2\int \frac{du}{1 + u^2} = -tan \frac{x}{2} + 2arctan \left[\left(tan \frac{x}{2}\right)\right]$$

$$\boxed{\int \frac{dx}{1 + \sec x} = -tan \frac{x}{2} + x + C}$$

154

$$\int \frac{\sec x}{\sec x + \tan x - 1}\, dx$$

The integral is rewritten

$$\int \frac{\sec x}{\sec x + \tan x - 1}\, dx = \int \frac{\dfrac{1}{\cos x}}{\dfrac{1}{\cos x} + \dfrac{sen\, x}{\cos x} - 1}\, dx$$

For the resolution of integrals containing sine and cosine functions, the following substitutions will be made.

$$sen\, x = \frac{2u}{1 + u^2}\quad,\quad \cos x = \frac{1 - u^2}{1 + u^2}\quad,\quad dx = \frac{2}{1 + u^2}\, du\ ,\quad u = \tan\frac{x}{2}$$

$$\int \frac{\sec x}{\sec x + \tan x - 1}\, dx = \int \frac{\dfrac{1}{\cos x}}{\dfrac{1}{\cos x} + \dfrac{sen\, x}{\cos x} - 1}\, dx$$

$$\int \frac{\sec x}{\sec x + \tan x - 1}\, dx = \int \frac{\dfrac{1}{\cos x}}{\dfrac{1 + sen\, x - \cos x}{\cos x}}\, dx$$

$$\int \frac{\sec x}{\sec x + \tan x - 1}\, dx = \int \frac{\cos x}{\cos x(1 + sen\, x - \cos x)}\, dx$$

$$\int \frac{\sec x}{\sec x + \tan x - 1}\, dx = \int \frac{dx}{1 + sen\, x - \cos x}$$

$$\int \frac{\sec x}{\sec x + \tan x - 1}\, dx = \int \frac{\dfrac{2}{1 + u^2}\, du}{1 + \dfrac{2u}{1 + u^2} - \dfrac{1 - u^2}{1 + u^2}}$$

$$\int \frac{\sec x}{\sec x + \tan x - 1} dx = \int \frac{\frac{2}{1 + u^2} du}{\frac{1 + u^2 + 2u - 1 + u^2}{1 + u^2}}$$

$$\int \frac{\sec x}{\sec x + \tan x - 1} dx = \int \frac{2}{2u(u + 1)} du = \int \frac{du}{u(u + 1)}$$

Where the resulting integral is solved by applying partial fraction integration.

$$\int \frac{du}{u(u + 1)} = \int \frac{A}{u} du + \int \frac{B}{u + 1} du$$

$$\frac{1}{u(u + 1)} = \frac{A}{u} + \frac{B}{u + 1} \qquad \rightarrow \qquad 1 = A(u + 1) + Bu$$

For $u = -1$ you have: $1 = -B \quad \rightarrow \quad B = -1$
For $u = 0$ you have: $1 = A \qquad \rightarrow \qquad A = 1$

$$\int \frac{du}{u(u + 1)} = \int \frac{du}{u} - \int \frac{du}{u + 1}$$

$$\int \frac{du}{u(u + 1)} = ln|u| - ln|u + 1| + C$$

As $u = tan\frac{x}{2}$

$$\int \frac{\sec x}{\sec x + \tan x - 1} dx = ln\left|tan\frac{x}{2}\right| - ln\left|tan\frac{x}{2} + 1\right| + C$$

Applying logarithm properties

$$\boxed{\int \frac{\sec x}{\sec x + \tan x - 1} dx = ln\left|\frac{tan\frac{x}{2}}{tan\frac{x}{2} + 1}\right| + C}$$

155

$$\int \frac{dx}{\tan x + sen\, x}$$

The integral is rewritten

$$\int \frac{dx}{\tan x + sen\, x} = \int \frac{dx}{\dfrac{sen\, x}{\cos x} + sen\, x}$$

For the resolution of integrals containing sine and cosine functions, the following substitutions will be made.

$$sen\, x = \frac{2u}{1 + u^2} \quad , \quad \cos x = \frac{1 - u^2}{1 + u^2} \quad , \quad dx = \frac{2}{1 + u^2}\, du \quad , \quad u = \tan\frac{x}{2}$$

The quotient of senx and cosx is initially resolved

$$\frac{sen\, x}{\cos x} = \frac{\dfrac{2u}{1 + u^2}}{\dfrac{1 - u^2}{1 + u^2}} = \frac{2u}{1 - u^2}$$

$$\int \frac{dx}{\dfrac{sen\, x}{\cos x} + sen\, x} = \int \frac{\dfrac{2}{1 + u^2}\, du}{\dfrac{2u}{1 - u^2} + \dfrac{2u}{1 + u^2}} = \int \frac{\dfrac{2}{1 + u^2}\, du}{\dfrac{2u(1 + u^2) + 2u(1 - u^2)}{(1 - u^2)(1 + u^2)}}$$

$$\int \frac{dx}{\dfrac{sen\, x}{\cos x} + sen\, x} = \int \frac{\dfrac{2}{1 + u^2}\, du}{\dfrac{2u + 2u^3 + 2u - 2u^3}{(1 - u^2)(1 + u^2)}} = \int \frac{\dfrac{2}{(1 + u^2)}\, du}{\dfrac{4u}{(1 - u^2)(1 + u^2)}}$$

$$\int \frac{dx}{\dfrac{sen\, x}{\cos x} + sen\, x} = \int \frac{2}{\dfrac{4u}{(1 - u^2)}}\, du = \int \frac{2 - 2u^2}{4u}\, du$$

$$\int \frac{dx}{\dfrac{sen\, x}{\cos x} + sen\, x} = \int \frac{2(1 - u^2)}{4u}\, du = \frac{1}{2}\int \frac{1 - u^2}{u}\, du$$

$$\int \frac{dx}{\frac{sen\,x}{cos\,x} + sen\,x} = -\frac{1}{2}\int \frac{u^2-1}{u}\,du \longrightarrow$$ It multiplied by minus (-) the integral to change position the terms of the numerator.

$$\int \frac{dx}{\frac{sen\,x}{cos\,x} + sen\,x} = -\frac{1}{2}\int \left(u-\frac{1}{u}\right)du \longrightarrow$$ The division of polynomials was made: $\frac{u^2-1}{u}$

$$\int \frac{dx}{\frac{sen\,x}{cos\,x} + sen\,x} = -\frac{1}{2}\int u\,du + \frac{1}{2}\int \frac{du}{u}$$

$$\int \frac{dx}{\frac{sen\,x}{cos\,x} + sen\,x} = -\frac{1}{4}u^2 + \frac{1}{2}ln|u| + C$$

$$\boxed{\int \frac{dx}{\frac{sen\,x}{cos\,x} + sen\,x} = \frac{1}{2}ln\left|tan\frac{x}{2}\right| - \frac{1}{4}tan^2\frac{x}{2} + C}$$

$$\boxed{\int \frac{sec\,x}{1+cos\,x}dx}$$ **156**

The integral is rewritten

$$\int \frac{sec\,x}{1+cos\,x}dx = \int \frac{\frac{1}{cos\,x}}{1+cos\,x}\,dx$$

For the resolution of integrals containing sine and cosine functions, the following substitutions will be made.

$$sen\,x = \frac{2u}{1+u^2}\quad,\quad cos\,x = \frac{1-u^2}{1+u^2}\quad,\quad dx = \frac{2}{1+u^2}du\quad,\quad u = tan\frac{x}{2}$$

It is solved initially $\frac{1}{cos\,x} : \frac{1}{cos\,x} = \frac{1}{\frac{1-u^2}{1+u^2}} = \frac{1+u^2}{1-u^2}$

$$\int \frac{\sec x}{1 + \cos x}\, dx = \int \frac{\dfrac{1+u^2}{1-u^2}\dfrac{2}{1+u^2}}{1 + \dfrac{1-u^2}{1+u^2}}\, du = \int \frac{\dfrac{1+u^2}{1-u^2}\dfrac{2}{1+u^2}}{\dfrac{1+u^2+1-u^2}{1+u^2}}\, du$$

$$\int \frac{\sec x}{1 + \cos x}\, dx = \int \frac{\dfrac{2+2u^2}{1-u^4}}{\dfrac{2}{1+u^2}}\, du = \int \frac{(2+2u^2)(1+u^2)}{2(1-u^4)}\, du$$

$$\int \frac{\sec x}{1 + \cos x}\, dx = \int \frac{2 + 2u^2 + 2u^2 + 2u^4}{2(1-u^4)}\, du$$

$$\int \frac{\sec x}{1 + \cos x}\, dx = \int \frac{2(1 + u^2 + u^2 + u^4)}{2(1-u^4)}\, du$$

$$\int \frac{\sec x}{1 + \cos x}\, dx = \frac{u^4 + 2u^2 + 1}{1 - u^4}$$

$$\int \frac{\sec x}{1 + \cos x}\, dx = \int \left(-1 + \frac{2u^2 + 2}{1 - u^4}\right) du$$

$$\int \frac{\sec x}{1 + \cos x}\, dx = -\int du + 2\int \frac{u^2 + 1}{1 - u^4}\, du$$

The denominator of the second integral is factored

$$1 - u^4 = 1^2 - (u^2)^2 = (1 + u^2)(1 - u^2)$$

Therefore, you have

$$\int \frac{\sec x}{1 + \cos x}\, dx = -\int du + 2\int \frac{u^2 + 1}{(1+u^2)(1-u^2)}\, du$$

$$\int \frac{\sec x}{1 + \cos x}\, dx = -\int du + 2\int \frac{du}{(1 - u^2)}$$

$$\int \frac{\sec x}{1 + \cos x}\, dx = -u + 2\frac{1}{2} \ln\left|\frac{1+u}{1-u}\right| + C$$

$$\boxed{\int \frac{\sec x}{1 + \cos x}\, dx = -\tan\frac{x}{2} + \ln\left|\frac{1 + \tan\frac{x}{2}}{1 - \tan\frac{x}{2}}\right| + C}$$

$$\int \frac{dx}{2 + sen\, x + \cos x}$$

157

For the resolution of integrals containing sine and cosine functions, the following substitutions will be made.

$$sen\, x = \frac{2u}{1 + u^2} \quad , \quad \cos x = \frac{1 - u^2}{1 + u^2} \quad , \quad dx = \frac{2}{1 + u^2}\, du \quad , \quad u = tan\frac{x}{2}$$

$$\int \frac{dx}{2 + sen\, x + \cos x} = \int \frac{\frac{2}{1 + u^2}\, du}{2 + \frac{2u}{1 + u^2} + \frac{1 - u^2}{1 + u^2}} = \int \frac{\frac{2}{1 + u^2}\, du}{\frac{2(1 + u^2) + 2u + 1 - u^2}{1 + u^2}}$$

$$\int \frac{dx}{2 + sen\, x + \cos x} = \int \frac{\frac{2}{1 + u^2}\, du}{\frac{2 + 2u^2 + 2u + 1 - u^2}{1 + u^2}} = \int \frac{\frac{2}{1 + u^2}\, du}{\frac{u^2 + 2u + 3}{1 + u^2}}$$

$$\int \frac{dx}{2 + sen\, x + \cos x} = 2\int \frac{du}{u^2 + 2u + 3} = 2\int \frac{du}{(u + 1)^2 + 2}$$

$$\int \frac{dx}{2 + sen\, x + \cos x} = 2\int \frac{du}{(u + 1)^2 + (\sqrt{2})^2} = 2\left(\frac{1}{\sqrt{2}} arctan\frac{u + 1}{\sqrt{2}}\right) + C$$

$$\int \frac{dx}{2 + sen\ x + \cos x} = \sqrt{2}arctan\left(\frac{tan\frac{x}{2} + 1}{\sqrt{2}}\right) + C$$

$$\boxed{\int \frac{dx}{2 + sen\ x + \cos x} = \sqrt{2}arctan\left[\frac{\sqrt{2}}{2}tan\frac{x}{2} + 1\right] + C}$$

158

$$\int \frac{dx}{4 - 5sen\ x}$$

$$\int \frac{dx}{4 - 5sen\ x} = \int \frac{\frac{2}{1 + u^2}du}{4 - 5\frac{2u}{1 + u^2}}$$

$$\int \frac{dx}{4 - 5sen\ x} = \int \frac{\frac{2}{1 + u^2}du}{\frac{4 + 4u^2 - 10u}{1 + u^2}}$$

$$\int \frac{dx}{4 - 5sen\ x} = 2\int \frac{du}{4u^2 - 10u + 4} \longrightarrow$$ Case of Factorization of the form:
$$ax^2 + bx + c = 0$$

$$\int \frac{dx}{4 - 5sen\ x} = 2\int \frac{du}{(u - 2)(u - \frac{1}{2})}$$

$$\int \frac{dx}{4 - 5sen\ x} = 2\int \frac{A}{u - 2}du + 2\int \frac{B}{u - \frac{1}{2}}du \rightarrow$$ Where: $A=\frac{2}{3}$ y $B=-\frac{2}{3}$

$$\int \frac{dx}{4 - 5sen\ x} = \frac{4}{3}\int \frac{dz}{z} - \frac{4}{3}\int \frac{dt}{t} \rightarrow$$ Where: $Z= u-2$, $dz = du$, $t = u - \frac{1}{2}$, $dt = du$

$$\int \frac{dx}{4 - 5sen\ x} = \frac{4}{3}ln|u - 2| - \frac{4}{3}ln\left|u - \frac{1}{2}\right| + C$$

$$\int \frac{dx}{4-5sen\,x} = \frac{4}{3} \ln \left| tan\left(\frac{x}{2}\right) - 2 \right| - \frac{4}{3} \ln \left| tan\left(\frac{x}{2}\right) - \frac{1}{2} \right| + C$$

$$\boxed{\int \frac{dx}{4-5sen\,x} = \frac{4}{3} \ln \left| \frac{tan\left(\frac{x}{2}\right) - 2}{2tan\left(\frac{x}{2}\right) - 1} \right| + C}$$

$$\boxed{\int \frac{dx}{\cos x + \cot x}}$$ 159

The integral is rewritten

$$\int \frac{dx}{\cos x + \cot x} = \int \frac{dx}{\cos x + \frac{\cos x}{sen\,x}}$$

For the resolution of integrals containing sine and cosine functions, the following substitutions will be made.

$$sen\,x = \frac{2u}{1+u^2} \quad , \quad \cos x = \frac{1-u^2}{1+u^2} \quad , \quad dx = \frac{2}{1+u^2}du \quad , \quad u = tan\frac{x}{2}$$

It solves the $\frac{\cos x}{sen\,x}$

$$\frac{\cos x}{sen\,x} = \frac{\frac{1-u^2}{1+u^2}}{\frac{2u}{1+u^2}} = \frac{1-u^2}{2u}$$

$$\int \frac{dx}{\cos x + \frac{\cos x}{sen\,x}} = \int \frac{\frac{2}{1+u^2}du}{\frac{1-u^2}{1+u^2} + \frac{1-u^2}{2u}} = \int \frac{\frac{2}{1+u^2}du}{\frac{2u(1-u^2)+(1+u^2)(1-u^2)}{2u(1+u^2)}}$$

$$\int \frac{dx}{\cos x + \frac{\cos x}{sen\,x}} = \int \frac{4u\,du}{(1-u^2)[2u+1+u^2]} = \int \frac{4u\,du}{(1-u)(1+u)[u^2+2u+1]}$$

$$\int \frac{dx}{\cos x + \dfrac{\cos x}{sen\,x}} = \int \frac{4u\,du}{(1+u)^3(1-u)} \quad \textbf{\textit{Equation 1}}$$

Now the resulting integral is solved by the method of partial fractions.

$$\int \frac{4u\,du}{(1+u)^3(1-u)} = \int \frac{A}{(1+u)} + \int \frac{B}{(1+u)^2} + \int \frac{C}{(1+u)^3} + \int \frac{D}{(1-u)}$$

$$\frac{4u}{(1+u)^3(1-u)} = \frac{A}{(1+u)} + \frac{B}{(1+u)^2} + \frac{C}{(1+u)^3} + \frac{D}{(1-u)}$$

$$4u = A(1+u)^2(1-u) + B(1+u)(1-u) + C(1-u) + D(1+u)^3$$

For u = 1 the factors that are made zero are those that contain a: A, B, and C.

so, 4 = 8D $\rightarrow$ $D = \dfrac{1}{2}$.

For u = -1 the factors that are made zero are those that contain a: A, B and D.

so, 2C = -4 $\rightarrow$ $C = -2$.

To find the values of A and B, the reader is asked to solve the fundamental equation whose result is

$$4u = (-A+D)u^3 + (-A-B+3D)u^2 + (A-C+3D)u + (A+B+C+D)$$

To find the values of A and B we have:

$$-A + D = 0 \rightarrow -A + \frac{1}{2} = 0 \rightarrow A = \frac{1}{2}$$

$$-A - B + 3D = -\frac{1}{2} - B + 3*\frac{1}{2} = 0 \quad \rightarrow \quad B = 1$$

Consequently, you have

$$\int \frac{4u\,du}{(1+u)^3(1-u)} = \int \frac{A}{(1+u)} + \int \frac{B}{(1+u)^2} + \int \frac{C}{(1+u)^3} + \int \frac{D}{(1-u)}$$

$$\int \frac{4u\,du}{(1+u)^3(1-u)} = \frac{1}{2}\int \frac{du}{(1+u)} + \int \frac{du}{(1+u)^2} - 2\int \frac{du}{(1+u)^3} + \frac{1}{2}\int \frac{du}{(1-u)}$$

$$\int \frac{4u\,du}{(1+u)^3(1-u)} = \frac{1}{2}ln\,|1+u| - \frac{1}{1+u} + \frac{1}{(1+u)^2} - \frac{1}{2}ln|1-u| + C$$

$$\int \frac{4u\,du}{(1+u)^3(1-u)} = \frac{1}{2}ln\,\left|\frac{1+u}{1-u}\right| - \frac{1}{1+u} + \frac{1}{(1+u)^2} + C$$

$$\int \frac{4u\,du}{(1+u)^3(1-u)} = \frac{1}{2}ln\,\left|\frac{1+u}{1-u}\right| + \frac{-(1+u)+1}{(1+u)^2} + C$$

$$\int \frac{4u\,du}{(1+u)^3(1-u)} = \frac{1}{2}ln\,\left|\frac{1+u}{1-u}\right| - \frac{u}{(1+u)^2} + C$$

$$\int \frac{4u\,du}{(1+u)^3(1-u)} = \frac{1}{2}ln\,\left|\frac{1+tan\frac{x}{2}}{1-tan\frac{x}{2}}\right| - \frac{tan\frac{x}{2}}{\left(1+tan\frac{x}{2}\right)^2} + C$$

Replace the previous result in equation 1 and obtain

$$\boxed{\int \frac{dx}{\cos x + \cot x} = \frac{1}{2}ln\,\left|\frac{1+tan\frac{x}{2}}{1-tan\frac{x}{2}}\right| - \frac{tan\frac{x}{2}}{\left(1+tan\frac{x}{2}\right)^2} + C}$$

$$\boxed{\int \frac{\csc x}{1 + sen\,x}dx}$$

160

The integral is rewritten

$$\int \frac{\csc x}{1 + sen\,x}dx = \int \frac{\frac{1}{sen\,x}}{1+sen\,x}dx = \int \frac{\frac{1}{2u}\frac{2}{1+u^2}du}{1+\frac{2u}{1+u^2}}$$

The numerator of the last integral is solved

$$\frac{1}{\dfrac{2u}{1+u^2}} \cdot \frac{2}{1+u^2} = \frac{1+u^2}{2u} \cdot \frac{2}{1+u^2} = \frac{1}{u}$$

$$\int \frac{\csc x}{1 + \operatorname{sen} x}\, dx = \int \frac{\dfrac{1}{u}\, du}{1 + \dfrac{2u}{1+u^2}} = \int \frac{\dfrac{1}{u}\, du}{\dfrac{u^2 + 2u + 1}{1+u^2}}$$

$$\int \frac{\csc x}{1 + \operatorname{sen} x}\, dx = \int \frac{u^2 + 1}{u(u^2 + 2u + 1)}\, du \quad \textbf{\textit{Equation 1}}$$

Now the resulting integral is solved by the method of partial fractions.

$$\int \frac{u^2 + 1}{u(u^2 + 2u + 1)}\, du = \int \frac{A}{u} + \int \frac{Bu + C}{u^2 + 2u + 1}$$

$$\frac{u^2 + 1}{u(u^2 + 2u + 1)} = \frac{A}{u} + \frac{Bu + C}{u^2 + 2u + 1}$$

$$u^2 + 1 = A(u^2 + 2u + 1) + (Bu + C)u$$

$$u^2 + 1 = Au^2 + 2Au + A + Bu^2 + Cu$$

$$u^2 + 1 = (A + B)u^2 + (2A + C)u + A$$

$$\left|
\begin{array}{lll}
A + B & & = 1 \\
2A & + C & = 0 \\
A & & = 1
\end{array}
\right.$$

The system of equations is solved and the following values are obtained

$$A = 1, \quad B = 0, \quad C = -2$$

Consequently, you have

$$\int \frac{u^2 + 1}{u(u^2 + 2u + 1)}\, du = \int \frac{A}{u} + \int \frac{Bu + C}{u^2 + 2u + 1}$$

$$\int \frac{u^2 + 1}{u(u^2 + 2u + 1)}\, du = \int \frac{du}{u} - 2 \int \frac{du}{u^2 + 2u + 1}$$

$$\int \frac{u^2 + 1}{u(u^2 + 2u + 1)}\, du = \int \frac{du}{u} - 2 \int \frac{du}{(u + 1)^2}$$

$$\int \frac{u^2 + 1}{u(u^2 + 2u + 1)}\, du = ln|u| + \frac{2}{u + 1} + C$$

The previous result is substituted in equation 1

$$\boxed{\int \frac{csc\, x}{1 + sen\, x}\, dx = ln\left|tan\left(\frac{x}{2}\right)\right| + \frac{2}{tan\left(\frac{x}{2}\right) + 1} + C}$$

$$\boxed{\int \frac{1 - senx}{(1 + senx)senx}\, dx} \qquad \text{161}$$

The integral is rewritten

$$\int \frac{1 - senx}{(1 + senx)senx}\, dx = \int \left[\frac{(1 - sen\, x)}{(1 + sen\, x)}\frac{1}{sen\, x}\right] dx$$

Proceed to divide $(1 - sen\, x)$ con $(1 + sen\, x)$ to separate the integral problem into two integrals

$$\int \left[\frac{(1 - sen\, x)}{(1 + sen\, x)}\frac{1}{sen\, x}\right] dx = \int \left[\left(1 - \frac{2\, sen\, x}{1 + sen\, x}\right)\frac{1}{sen\, x}\right] dx$$

$$\int \left[\frac{(1 - sen\, x)}{(1 + sen\, x)}\frac{1}{sen\, x}\right] dx = \int \frac{dx}{sen\, x} - 2 \int \left(\frac{sen\, x}{1 + sen\, x}\right)\frac{1}{sen\, x}\, dx$$

$$\int \left[\frac{(1 - sen\, x)}{(1 + sen\, x)}\frac{1}{sen\, x}\right] dx = \int \frac{dx}{sen\, x} - 2 \int \frac{dx}{1 + sen\, x} \qquad Ec.\, 1$$

The first and second integral are solved

$$\int \frac{dx}{sen\ x} = \int \frac{\frac{2du}{1+u^2}}{\frac{2u}{1+u^2}} = \int \frac{du}{u} = ln|u| = ln\left|tan\frac{x}{2}\right| + C_1$$

$$2\int \frac{dx}{1+sen\ x} = 2\int \frac{\frac{2du}{1+u^2}}{1+\frac{2u}{1+u^2}} = 2\int \frac{\frac{2du}{1+u^2}}{\frac{u^2+2u+1}{1+u^2}} = 4\int (u+1)^{-2}\ du$$

$$2\int \frac{dx}{1+sen\ x} = -\frac{4}{tan\left(\frac{x}{2}\right)+1} + C_2$$

Substituting these results in equation 1 you get

$$\boxed{\int \frac{1-senx}{(1+senx)senx}dx = ln\left|tan\frac{x}{2}\right| + \frac{4}{tan\left(\frac{x}{2}\right)+1} + C}$$

162

$$\boxed{\int \frac{sec\ x\ dx}{5\tan x + 3\ sec\ x + 3}}$$

Initial work is done making the respective substitutions in the denominator

$$5\tan x + 3\ sec\ x + 3 = \frac{5senx}{cosx} + \frac{3}{cos\ x} + 3 = \frac{\frac{10\ u}{1+u^2}}{\frac{1-u^2}{1+u^2}} + \frac{3}{\frac{1-u^2}{1+u^2}} + 3$$

$$5\tan x + 3\ sec\ x + 3 = \frac{10u}{1-u^2} + \frac{3+3u^2}{1-u^2} + 3 = \frac{10u+3+3u^2+3-3u^2}{1-u^2} = \frac{10u+6}{1-u^2}$$

Then with the numerator

$$secx\ dx = \frac{1}{cosx}dx = \frac{1}{\frac{1-u^2}{1+u^2}}\frac{2\ du}{1+u^2} = \frac{2\ du}{1-u^2}$$

Consequently, you have

$$\int \frac{\sec x \; dx}{5 \tan x + 3 \sec x + 3} = \int \frac{\frac{2 \, du}{1 - u^2}}{\frac{10u + 6}{1 - u^2}} = 2 \int \frac{du}{10u + 6} = \frac{1}{5} \int \frac{dz}{z}$$

$$\int \frac{\sec x \; dx}{5 \tan x + 3 \sec x + 3} = \frac{1}{5} \ln|z| + C = \frac{1}{5} \ln|10 \, u + 6| + C$$

$$\int \frac{\sec x \; dx}{5 \tan x + 3 \sec x + 3} = \frac{1}{5} \ln|2(5u + 3)| = \frac{1}{5} \ln|5u + 3| + \ln 2 + C$$

$$\int \frac{\sec x \; dx}{5 \tan x + 3 \sec x + 3} = \frac{1}{5} \ln \left| 5 \tan \left(\frac{x}{2} \right) + 3 \right| + C$$

$$\boxed{\int \frac{\sec x \; dx}{5 \tan x + 3 \sec x + 3} = \frac{1}{5} \ln \left| 5 \tan \left(\frac{x}{2} \right) + 3 \right| + C}$$

$$\boxed{\int \frac{dx}{\cos^2 x + 5 \cos x + 6}}$$

163

The denominator is factored

$$\cos^2 x + 5 \cos x + 6 = (\cos x + 3)(\cos x + 2)$$

$$\int \frac{dx}{\cos^2 x + 5 \cos x + 6} = \int \frac{dx}{(\cos x + 3)(\cos x + 2)}$$

To solve the integral, the partial fractions integration method is applied

$$\int \frac{dx}{(\cos x + 3)(\cos x + 2)} = \int \frac{A \; dx}{\cos x + 3} + \int \frac{B \; dx}{\cos x + 2}$$

$$1 = A(\cos x + 2) + B(\cos x + 3)$$

$$1 = A \cos x + 2A + B \cos x + 3B$$

$$1 = (A + B) \cos x + (2A + 3B)$$

$$
\begin{aligned}
A + B &= 0 \\
2A + 3B &= 1
\end{aligned}
\qquad \text{Multiply the first equation by -2}
$$

$$\downarrow$$

$$
\begin{aligned}
-2A - 2B &= 0 \\
2A + 3B &= 1
\end{aligned}
\qquad \rightarrow \quad B = 1 \qquad A = -1
$$

$$
\int \frac{dx}{(\cos x + 3)(\cos x + 2)} = \int \frac{A\,dx}{\cos x + 3} + \int \frac{B\,dx}{\cos x + 2}
$$

$$
\int \frac{dx}{\cos^2 x + 5\cos x + 6} = -\int \frac{dx}{\cos x + 3} + \int \frac{dx}{\cos x + 2} \qquad \textbf{\textit{Equation 1}}
$$

The integral is solved $\displaystyle\int \frac{dx}{\cos x + 3}$

$$
\int \frac{dx}{\cos x + 3} = \int \frac{\dfrac{2du}{1 + u^2}}{\dfrac{1 - u^2}{1 + u^2} + 3}
$$

$$
= \int \frac{\dfrac{2du}{1 + u^2}}{\dfrac{1 - u^2 + 3 + 3u^2}{1 + u^2}}
$$

$$
= 2 \int \frac{du}{2u^2 + 4}
$$

$$
= 2 \int \frac{du}{(\sqrt{2}\,u)^2 + 2^2}
$$

$$
= \arctan\left(\frac{\sqrt{2}\,u}{2}\right) + C_1
$$

$$\int \frac{dx}{\cos x + 3} = -\arctan\left[\frac{1}{2}\tan\left(\frac{x}{2}\right)\sqrt{2}\right] + C_1$$

Solve the second integral of equation 1

$$\int \frac{dx}{\cos x + 2} = \int \frac{\frac{2du}{1+u^2}}{\frac{1-u^2}{1+u^2} + 2}$$

$$= \int \frac{\frac{2du}{1+u^2}}{\frac{1-u^2+2+2u^2}{1+u^2}}$$

$$= 2\int \frac{du}{u^2 + 3}$$

$$= 2\int \frac{du}{u^2 + (\sqrt{3})^2}$$

$$\int \frac{dx}{\cos x + 2} = \frac{2}{\sqrt{3}}\arctan\left(\frac{u}{\sqrt{3}}\right) + C_2$$

$$\int \frac{dx}{\cos x + 2} = \frac{2}{3}\sqrt{3}\arctan\left[\frac{1}{3}\tan\left(\frac{x}{2}\right)\sqrt{3}\right] + C_2$$

The results of the integrals of equation 1 are substituted

$$\int \frac{dx}{\cos^2 x + 5\cos x + 6} = -\arctan\left[\frac{1}{2}\tan\left(\frac{x}{2}\right)\sqrt{2}\right] + \frac{2}{3}\sqrt{3}\arctan\left[\frac{1}{3}\tan\left(\frac{x}{2}\right)\sqrt{3}\right] + C$$

164

$$\int \frac{dx}{3sen\,x + 4\cos x}$$

The respective substitutions are made.

$$\int \frac{dx}{3sen\,x + 4\cos x} = \int \frac{\dfrac{2du}{1+u^2}}{3\dfrac{2u}{1+u^2} + 4\dfrac{1-u^2}{1+u^2}} = \int \frac{\dfrac{2du}{1+u^2}}{\dfrac{6u}{1+u^2} + \dfrac{4-4u^2}{1+u^2}}$$

$$= \int \frac{\dfrac{2du}{1+u^2}}{\dfrac{6u+4-4u^2}{1+u^2}} = -2\int \frac{du}{4u^2 - 6u - 4}$$

The denominator is factored and the integral is solved by applying partial fractions

$$4u^2 - 6u - 4 = 16u^2 - 6(4u) - 16 = (4u-8)(4u+2) = (u-2)(4u+2)$$

$$\int \frac{dx}{3sen\,x + 4\cos x} = -2\int \frac{du}{(u-2)(4u+2)}$$

$$\int \frac{dx}{3sen\,x + 4\cos x} = -2\int \frac{A\,du}{(u-2)} - 2\int \frac{B\,du}{(4u+2)}$$

$$1 = A(4u+2) + B(u-2)$$

$$1 = 4Au + 2A + Bu - 2B$$

$$1 = (4A+B)\,u + (2A-B)$$

$$\begin{array}{ll} 4A + B = 0 & \rightarrow A = \dfrac{1}{10} \quad B = -\dfrac{2}{5} \\ 2A - 2B = 1 & \end{array}$$

$$-2\int \frac{A\,du}{(u-2)} - 2\int \frac{B\,du}{(4u+2)} = -\frac{1}{5}\int \frac{du}{(u-2)} + \frac{4}{5}\int \frac{du}{(4u+2)}$$

$$= -\frac{1}{5}\ln|u - 2| + \frac{4}{20}\ln|4u + 2| + C$$

$$= -\frac{1}{5}\ln\left|\tan\left(\frac{x}{2}\right) - 2\right| + \frac{1}{5}\ln\left|4\tan\left(\frac{x}{2}\right) + 2\right| + C$$

$$= -\frac{1}{5}\ln\left|\tan\left(\frac{x}{2}\right) - 2\right| + \frac{1}{5}\ln\left|4\left[\tan\left(\frac{x}{2}\right) + \frac{2}{4}\right]\right| + C$$

$$= -\frac{1}{5}\ln\left|\tan\left(\frac{x}{2}\right) - 2\right| + \frac{1}{5}\ln\left|\tan\left(\frac{x}{2}\right) + \frac{1}{2}\right| + C$$

$$\boxed{\int \frac{dx}{3\,sen\,x + 4\cos x} = \frac{1}{5}\ln\left|\frac{\tan\left(\frac{x}{2}\right) + \frac{1}{2}}{\tan\left(\frac{x}{2}\right) - 2}\right| + C}$$

Note: In the fourth step the 4 leaves the absolute value as ln 4 and is added to the constant (C).

$$\boxed{\int \frac{dx}{1 + sen\,x - \cos x}}$$ **165**

$$\int \frac{dx}{1 + sen\,x - \cos x} = \int \frac{\dfrac{2du}{1 + u^2}}{1 + \dfrac{2u}{1 + u^2} - \dfrac{1 - u^2}{1 + u^2}}$$

$$= \int \frac{\dfrac{2du}{1 + u^2}}{\dfrac{1 + u^2 + 2u - 1 + u^2}{1 + u^2}}$$

$$\int \frac{dx}{1 + sen\,x - \cos x} = 2\int \frac{du}{2u^2 + 2u} = \int \frac{du}{u(u + 1)} = \int \frac{A\,du}{u} + \int \frac{B\,du}{u + 1}$$

$$1 = A(u + 1) + B(u)$$

$$1 = Au + A + Bu$$

$$1 = (A + B)u + A$$

$$A + B = 0 \qquad \rightarrow \quad A = 1 \ and \ B = -1$$
$$A = 1$$

$$\int \frac{du}{u(u+1)} = \int \frac{A \, du}{u} + \int \frac{B \, du}{u+1} = \int \frac{du}{u} - \int \frac{du}{u+1} = ln|u| - ln|u+1| + C$$

$$\boxed{\int \frac{dx}{1 + sen \ x - cos \ x} = ln \left| \frac{tan\left(\frac{x}{2}\right)}{1 + tan\left(\frac{x}{2}\right)} \right| + C}$$

166

$$\boxed{\int \frac{dx}{1 - 2sen \ x}}$$

$$\int \frac{dx}{1 - 2sen \ x} = \int \frac{\dfrac{2du}{1+u^2}}{1 - 2\dfrac{2u}{1+u^2}} = \int \frac{\dfrac{2du}{1+u^2}}{1 - \dfrac{4u}{1+u^2}}$$

$$= \int \frac{\dfrac{2du}{1+u^2}}{\dfrac{1+u^2-4u}{1+u^2}} = \int \frac{\dfrac{2du}{1+u^2}}{\dfrac{u^2-4u+1}{1+u^2}}$$

$$\int \frac{dx}{1 - 2sen \ x} = 2 \int \frac{du}{u^2 - 4u + 1}$$

Complete squares to get an immediate integral

$$u^2 - 4u + 1 = (u^2 - 4u + 4) + 1 - 4$$

$$u^2 - 4u + 1 = (u - 2)^2 - 3$$

$$u^2 - 4u + 1 = (u - 2)^2 - (\sqrt{3})^2$$

$$\int \frac{dx}{1 - 2sen\ x} = 2\int \frac{du}{u^2 - 4u + 1} = 2\int \frac{du}{(u-2)^2 - (\sqrt{3})^2}$$

$$\int \frac{dx}{1 - 2sen\ x} = 2\frac{1}{2\sqrt{3}}\ ln\left|\frac{(u-2)-\sqrt{3}}{(u-2)+\sqrt{3}}\right| + C$$

$$\boxed{\int \frac{dx}{1 - 2sen\ x} = \frac{\sqrt{3}}{3}\ ln\left|\frac{tan\left(\frac{x}{2}\right) - 2 - \sqrt{3}}{tan\left(\frac{x}{2}\right) - 2 + \sqrt{3}}\right| + C}$$

$$\boxed{\int \frac{dx}{5 + 4sen\ x}}$$ **167**

$$\int \frac{dx}{5 + 4sen\ x} = \int \frac{\dfrac{2du}{1+u^2}}{5 + 4\dfrac{2u}{1+u^2}} = \int \frac{\dfrac{2du}{1+u^2}}{5 + \dfrac{8u}{1+u^2}}$$

$$= \int \frac{\dfrac{2du}{1+u^2}}{\dfrac{5 + 5u^2 + 8u}{1+u^2}} = \int \frac{\dfrac{2du}{1+u^2}}{\dfrac{5u^2 + 8u + 5}{1+u^2}}$$

$$\int \frac{dx}{5 + 4sen\ x} = 2\int \frac{du}{5u^2 + 8u + 5}$$

Complete squares to get an immediate integral

$$5u^2 + 8u + 5 = 5\left(u^2 + \frac{8}{5}u + 1\right)$$

$$5u^2 + 8u + 5 = 5\left(u^2 + \frac{8}{5}u + \frac{16}{25}\right) + 1 - \frac{16}{25}$$

$$5u^2 + 8u + 5 = 5(u + \frac{4}{5})^2 + 9$$

$$5u^2 + 8u + 5 = 5(u + \frac{4}{5})^2 + 3^2$$

$$\int \frac{dx}{5 + 4 sen\ x} = 2\int \frac{du}{5u^2 + 8u + 5} = 2\int \frac{du}{5(u + \frac{4}{5})^2 + 3^2}$$

$$\boxed{\int \frac{dx}{5 + 4 sen\ x} = \frac{2}{3} tan^{-1}\left(\frac{5 tan\left(\frac{x}{2}\right) + 4}{3}\right) + C}$$

168

$$\int \frac{dx}{1 - 2 sen\ x}$$

$$\int \frac{dx}{1 + sen\ x} = \int \frac{\frac{2du}{1 + u^2}}{1 + \frac{2u}{1 + u^2}} = \int \frac{\frac{2du}{1 + u^2}}{\frac{1 + u^2 + 2u}{1 + u^2}}$$

$$= 2\int \frac{du}{u^2 + 2u + 1} = 2\int \frac{du}{(u + 1)^2} = 2\int (u + 1)^{-2} du$$

$$\boxed{\int \frac{dx}{1 + sen\ x} = -\frac{2}{tan\left(\frac{x}{2}\right) + 1} + C}$$

169

$$\int \frac{dx}{5 + 3\cos x}$$

$$\int \frac{dx}{5 + 3\cos x} = \int \frac{\frac{2du}{1 + u^2}}{5 + 3\frac{1 - u^2}{1 + u^2}} = \int \frac{\frac{2du}{1 + u^2}}{\frac{5 + 5u^2 + 3 - 3u^2}{1 + u^2}}$$

$$\int \frac{dx}{5 + 3\cos x} = \int \frac{2du}{2u^2 + 8} = 2\int \frac{du}{2(u^2 + 4)}$$

$$\int \frac{dx}{5 + 3\cos x} = \int \frac{du}{(u^2 + 4)} = \int \frac{du}{(u^2 + 2^2)}$$

$$\boxed{\int \frac{dx}{5 + 3\cos x} = \frac{1}{2}\tan^{-1}\left(\frac{\tan\left(\frac{x}{2}\right)}{2}\right) + C}$$

$$\boxed{\int \frac{1}{2 - 3sen\,\theta}\,d\theta}$$

170

$$\int \frac{1}{2 - 3sen\,\theta}\,d\theta = \int \frac{\dfrac{2du}{1 + u^2}}{2 - 3\dfrac{2u}{1 + u^2}} = \int \frac{\dfrac{2du}{1 + u^2}}{2 - \dfrac{6u}{1 + u^2}}$$

$$\int \frac{1}{2 - 3sen\,\theta}\,d\theta = \int \frac{\dfrac{2du}{1 + u^2}}{\dfrac{2 + 2u^2 - 6u}{1 + u^2}}$$

$$\int \frac{1}{2 - 3sen\,\theta}\,d\theta = 2\int \frac{du}{2u^2 - 6u + 2} = \int \frac{du}{u^2 - 3u + 1}$$

Complete squares to get an immediate integral

$$u^2 - 3u + 1 = \left(u^2 - 3u + \frac{9}{4}\right) + 1 - \frac{9}{4}$$

$$u^2 - 3u + 1 = \left(u - \frac{3}{2}\right)^2 - \frac{5}{4}$$

$$u^2 - 3u + 1 = \left(u - \frac{3}{2}\right)^2 - \left(\frac{\sqrt{5}}{\sqrt{4}}\right)^2$$

$$\int \frac{du}{u^2 - 3u + 1} = \int \frac{du}{\left(u - \frac{3}{2}\right)^2 - \left(\frac{\sqrt{5}}{\sqrt{4}}\right)^2} = \frac{1}{2\frac{\sqrt{5}}{2}} ln \left| \frac{u - \frac{3}{2} - \frac{\sqrt{5}}{2}}{u - \frac{3}{2} + \frac{\sqrt{5}}{2}} \right| + C$$

$$\boxed{\int \frac{1}{2 - 3 sen\,\theta}\, d\theta = \frac{\sqrt{5}}{5} ln \left| \frac{2 tan\left(\frac{\theta}{2}\right) - 3 - \sqrt{5}}{2 tan\left(\frac{\theta}{2}\right) - 3 + \sqrt{5}} \right| + C}$$

171

$$\int \frac{1}{1 + sen\,\theta + \cos\theta}\, d\theta$$

$$\int \frac{1}{1 + sen\,\theta + \cos\theta}\, d\theta = \int \frac{\frac{2du}{1 + u^2}}{1 + \frac{2u}{1 + u^2} + \frac{1 - u^2}{1 + u^2}} = \int \frac{\frac{2du}{1 + u^2}}{\frac{1 + u^2 + 2u + 1 - u^2}{1 + u^2}}$$

$$\int \frac{1}{1 + sen\,\theta + \cos\theta}\, d\theta = 2 \int \frac{du}{2u + 2} = \frac{2}{2} \int \frac{du}{u + 1} = ln|u + 1| + C$$

$$\boxed{\int \frac{1}{1 + sen\,\theta + \cos\theta}\, d\theta = ln\left(tan\left(\frac{x}{2}\right) + 1\right) + C}$$

172

$$\int \frac{sen\,\theta}{1 + \cos^2\theta}\, d\theta$$

The integral is rewritten

$$\int \frac{sen\,\theta}{1 + \cos^2\theta}\, d\theta = \int \frac{sen\,\theta}{1 + (\cos\theta)^2}\, d\theta$$

It does: $u = \cos\theta \quad \rightarrow \quad du = -sen\,\theta\, d\theta$

$$\int \frac{sen\,\theta}{1+(\cos\theta)^2}\,d\theta = -\int \frac{-sen\,\theta}{1+(\cos\theta)^2}\,d\theta = -\int \frac{du}{1+u^2}$$

$$\boxed{\int \frac{sen\,\theta}{1+\cos^2\theta}\,d\theta = -tan^{-1}(\cos\theta)+C}$$

$$\boxed{\int \frac{sen\,\theta}{3-2\cos\theta}\,d\theta} \qquad \blacksquare\ 173$$

Before applying the respective sine, cosine and differential substitutions, first analyze if there is an alternative method for solving the integral. For example, try the substitution of u and du.

$$u = 3 - 2\cos\theta \quad \rightarrow \quad du = 2sen\,\theta\,d\theta$$

To obtain 2sen θ dθ in the integrand, simply multiply and divide the integral by 2

$$\frac{1}{2}\int \frac{2sen\,\theta}{3-2\cos\theta}\,d\theta = \int \frac{du}{u} = \frac{1}{2}ln|u|+C = \frac{1}{2}ln|3-2\cos\theta|+C$$

$$\boxed{\int \frac{sen\,\theta}{3-2\cos\theta}\,d\theta = \frac{1}{2}ln(3-2\cos\theta)+C}$$

$$\boxed{\int \frac{1}{3-2\cos\theta}\,d\theta} \qquad \blacksquare\ 174$$

$$\int \frac{1}{3-2\cos\theta}\,d\theta = \int \frac{\dfrac{2du}{1+u^2}}{3-2\dfrac{1-u^2}{1+u^2}} = \int \frac{\dfrac{2du}{1+u^2}}{\dfrac{3+3u^2-2+2u^2}{1+u^2}}$$

$$\int \frac{1}{3-2\cos\theta}\,d\theta = 2\int \frac{du}{5u^2+1} = 2\int \frac{du}{(\sqrt{5}u)^2+1}$$

It does $z = \sqrt{5}u \longrightarrow dz = \sqrt{5}du$

$$2\int \frac{du}{(\sqrt{5}u)^2 + 1} = \frac{2}{\sqrt{5}}\int \frac{dz}{z^2 + 1^2} = \frac{2}{\sqrt{5}}\arctan(z) + C$$

$$\boxed{\int \frac{1}{3 - 2\cos\theta}\,d\theta = \frac{2\sqrt{5}}{5}\,arctan\left(\sqrt{5}\,tan\frac{\theta}{2}\right) + C}$$

175 $\qquad\qquad\qquad\qquad\qquad\qquad\qquad\qquad\boxed{\int \frac{\cos\theta}{1 + \cos\theta}\,d\theta}$

Before applying the respective cosine and differential substitutions, first analyze if there is an alternative method for solving the integral. For example, multiply and divide by the conjugate of the denominator.

$$\int \frac{\cos\theta}{1 + \cos\theta}\,d\theta = \int \frac{(\cos\theta)(1 - \cos\theta)}{(1 + \cos\theta)(1 - \cos\theta)}\,d\theta$$

$$\int \frac{\cos\theta}{1 + \cos\theta}\,d\theta = \int \frac{\cos\theta - \cos^2\theta}{1 - \cos^2\theta}\,d\theta = \int \frac{\cos\theta - \cos^2\theta}{sen^2\theta}\,d\theta$$

$$\int \frac{\cos\theta}{1 + \cos\theta}\,d\theta = \int \frac{cos\theta}{sen^2\theta}\,d\theta - \int \frac{\cos^2\theta}{sen^2\theta}\,d\theta$$

$$\int \frac{\cos\theta}{1 + \cos\theta}\,d\theta = \int \frac{cos\theta}{sen\theta}\,\frac{1}{sen\theta} - \int \frac{\cos^2\theta}{sen^2\theta}\,d\theta$$

$$\int \frac{\cos\theta}{1 + \cos\theta}\,d\theta = \int csc\theta\,cot\theta\,d\theta - \int cot^2\theta\,d\theta$$

$$\int \frac{\cos\theta}{1+\cos\theta}\,d\theta = \int \csc\theta\,\cot\theta\,d\theta - \int (\csc^2\theta - 1)\,d\theta$$

$$\int \frac{\cos\theta}{1+\cos\theta}\,d\theta = \int \csc\theta\,\cot\theta\,d\theta - \int \csc^2\theta\,d\theta + \int d\theta$$

$$\int \frac{\cos\theta}{1+\cos\theta}\,d\theta = -\csc\theta + \cot\theta + \theta + C$$

$$\boxed{\int \frac{\cos\theta}{1+\cos\theta}\,d\theta = -\csc\theta + \cot\theta + \theta + C}$$

In the following exercises you will find the integrals of quadratic functions

$$\int \frac{dx}{x^2 - 4x + 13}$$

176

It completes squares

$$x^2 - 4x + 13 = (x^2 - 4x + 4) + 13 - 4 = (x - 2)^2 + 9$$

$$\int \frac{dx}{x^2 - 4x + 13} = \int \frac{dx}{(x - 2)^2 + 3^2} \quad \rightarrow \quad u = x - 2 \rightarrow du = dx\,, a = 3$$

$$\int \frac{dx}{u^2 + a^2} = \frac{1}{3} tan^{-1}\left(\frac{x - 2}{3}\right) + C$$

$$\boxed{\int \frac{dx}{x^2 - 4x + 13} = \frac{1}{3} tan^{-1}\left(\frac{x - 2}{3}\right) + C}$$

$$\int \frac{dx}{\sqrt{-x^2 - 4x - 2}}$$

177

It completes squares

$$-x^2 - 4x - 2 = -[x^2 + 4x + 2] = -[(x^2 + 4x + 4) + 2 - 4]$$

$$-x^2 - 4x - 2 = -[(x + 2)^2 - 2] = 2 - (x + 2)^2 = (\sqrt{2})^2 - (x + 2)^2$$

$$\int \frac{dx}{\sqrt{-x^2 - 4x - 2}} = \int \frac{dx}{\sqrt{\left(\sqrt{2}\right)^2 - (x + 2)^2}} \quad \rightarrow \quad u = x + 2, du = dx, a = \sqrt{2}$$

$$\int \frac{dx}{\sqrt{\left(\sqrt{2}\right)^2 - (x + 2)^2}} = \int \frac{dx}{\sqrt{a^2 - u^2}} = sen^{-1}\left(\frac{x + 2}{\sqrt{2}}\right) + C$$

$$\boxed{\int \frac{dx}{\sqrt{-x^2 - 4x - 2}} = sen^{-1}\left(\frac{x + 2}{\sqrt{2}}\right) + C}$$

178

$$\int \frac{2x+7}{x^2+2x+5}\,dx$$

The substitution or change of variable u and du is made with the intention of simplifying the integral problem

$$u = x^2 + 2x + 5 \longrightarrow du = (2x+2)dx$$

Observe the reader that appears in the differential (2x + 2). Through an artifice, the numerator of the integral is rewritten in such a way that it appears (2x + 2). And then the integral is separated into two integrals.

$$\int \frac{2x+7}{x^2+2x+5}\,dx = \int \frac{2x+2+5}{x^2+2x+5}\,dx = \int \frac{(2x+2)dx}{x^2+2x+5} + 5\int \frac{dx}{x^2+2x+5}$$

$$\int \frac{2x+7}{x^2+2x+5}\,dx = \int \frac{du}{u} + 5\int \frac{dx}{x^2+2x+5}$$

You fill squares in the second integral, and then you get the final result.

$$x^2 + 2x + 5 = x^2 + 2x + 1 + 5 - 1 = (x+1)^2 + 4$$

$$\int \frac{2x+7}{x^2+2x+5}\,dx = \int \frac{du}{u} + 5\int \frac{dx}{x^2+2x+5} = \int \frac{du}{u} + 5\int \frac{dx}{(x+1)^2+2^2}$$

$$z = x+1, \qquad dz = dx, \qquad a = 2$$

$$\int \frac{2x+7}{x^2+2x+5}\,dx = \int \frac{du}{u} + 5\int \frac{dz}{z^2+a^2}$$

$$\int \frac{2x+7}{x^2+2x+5}\,dx = \ln|x^2+2x+5| + \frac{5}{2}tan^{-1}\left(\frac{x+1}{2}\right) + C$$

$$\boxed{\int \frac{2x+7}{x^2+2x+5}\,dx = \ln(x^2+2x+5) + \frac{5}{2}tan^{-1}\left(\frac{1}{2}x+\frac{1}{2}\right) + C}$$

$$\int \frac{dx}{(x-1)\sqrt{x^2-2x-8}}$$

179

The substitution or change of variable u and du is made with the intention of simplifying the integral problem. We proceed also to complete squares.

$$u = x - 1 \;\; \longrightarrow \;\; du = dx$$

$$\int \frac{dx}{(x-1)\sqrt{x^2-2x-8}} = \int \frac{du}{u\sqrt{x^2-2x-8}}$$

It completes squares

$$x^2 - 2x - 8 = (x^2 - 2x + 1) - 8 - 1 = (x-1)^2 - 3^2$$

$$\int \frac{dx}{(x-1)\sqrt{x^2-2x-8}} = \int \frac{du}{u\sqrt{(x-1)^2-3^2}} = \int \frac{du}{u\sqrt{u^2-a^2}}$$

Through the two previous procedures, an immediate integral was obtained. And it is a reverse secant.

$$\int \frac{dx}{(x-1)\sqrt{x^2-2x-8}} = \frac{1}{3}sec^{-1}\left(\frac{x-1}{3}\right) + C$$

$$\boxed{\int \frac{dx}{(x-1)\sqrt{x^2-2x-8}} = \frac{1}{3}sec^{-1}\left(\frac{x-1}{3}\right) + C}$$

$$\int \frac{2x+5}{\sqrt{16-6x-x^2}}dx$$

180

$$u = 16 - 6x - x^2, \;\; du = (-6 - 2x)dx$$

Through the two previous procedures, an immediate integral was obtained. And it is a reverse secant.

$$\int \frac{2x+5}{\sqrt{16-6x-x^2}}dx = -\int \frac{-(2x+5)}{\sqrt{16-6x-x^2}}dx = -\int \frac{-2x-5-1+1}{\sqrt{16-6x-x^2}}dx$$

$$\int \frac{2x+5}{\sqrt{16-6x-x^2}}\,dx = -\int \frac{(-2x-6)\,dx}{\sqrt{16-6x-x^2}} - \int \frac{dx}{\sqrt{16-6x-x^2}}$$

You fill squares in the second integral, and then you get the final result.

$$16-6x-x^2 = -[x^2+6x-16] = -[(x^2+6x+9)-16-9]$$

$$16-6x-x^2 = -[(x+3)^2-25] = [25-(x+3)^2] = [5^2-(x+3)^2]$$

$$\int \frac{2x+5}{\sqrt{16-6x-x^2}}\,dx = -\int u^{-\frac{1}{2}}du - \int \frac{dz}{\sqrt{a^2-z^2}}$$

$$\int \frac{2x+5}{\sqrt{16-6x-x^2}}\,dx = -2\sqrt{16-6x-x^2} - sen^{-1}\left(\frac{x+3}{5}\right) + C$$

$$\boxed{\int \frac{2x+5}{\sqrt{16-6x-x^2}}\,dx = -2\sqrt{16-6x-x^2} - sen^{-1}\left(\frac{x+3}{5}\right) + C}$$

181

$$\boxed{\int \frac{6x-1}{4x^2+4x+10}\,dx}$$

The integral is rewritten   **Equation 1**

$$\int \frac{6x-1}{4x^2+4x+10}\,dx = 6\int \frac{x}{4x^2+4x+10}\,dx - \int \frac{dx}{4x^2+4x+10}$$

The first integral is solved

$$6\int \frac{x}{4x^2+4x+10}\,dx \quad , \quad u = 4x^2+4x+10 \quad , \quad du = (8x+4)dx$$

Multiply and divide the integral by 8 and add and subtract 4

$$6\int \frac{x}{4x^2+4x+10}\,dx = \frac{6}{8}\int \frac{(8x+4-4)dx}{4x^2+4x+10}$$

$$6\int \frac{x}{4x^2+4x+10}\,dx = \frac{3}{4}\int \frac{(8x+4)dx}{4x^2+4x+10} - 3\int \frac{dx}{4x^2+4x+10}$$

This result is substituted in equation 1 and is operated mathematically

$$\int \frac{6x - 1}{4x^2 + 4x + 10} dx = \frac{3}{4} \int \frac{(8x + 4)dx}{4x^2 + 4x + 10} - 3 \int \frac{dx}{4x^2 + 4x + 10} - \int \frac{dx}{4x^2 + 4x + 10}$$

Equation 2

$$\int \frac{6x - 1}{4x^2 + 4x + 10} dx = \frac{3}{4} \int \frac{(8x + 4)dx}{4x^2 + 4x + 10} - 4 \int \frac{dx}{4x^2 + 4x + 10}$$

Solve the second integral of equation 2

$$4 \int \frac{dx}{4x^2 + 4x + 10}$$

It completes squares

$$4x^2 + 4x + 10 = 4\left[x^2 + x + \frac{10}{4}\right] = 4\left[(x^2 + x + \frac{1}{4}) + \frac{10}{4} - \frac{1}{4}\right]$$

$$4x^2 + 4x + 10 = 4\left[\left(x + \frac{1}{2}\right)^2 + \frac{9}{4}\right] = 4\left[\left(x + \frac{1}{2}\right)^2 + \left(\frac{3}{2}\right)^2\right]$$

$$4 \int \frac{dx}{4x^2 + 4x + 10} = \frac{4}{4} \int \frac{dx}{\left(x + \frac{1}{2}\right)^2 + \left(\frac{3}{2}\right)^2} = \int \frac{dx}{\left(x + \frac{1}{2}\right)^2 + \left(\frac{3}{2}\right)^2}$$

This result is substituted in equation 2, and then the final result is obtained.

$$\int \frac{6x - 1}{4x^2 + 4x + 10} dx = \frac{3}{4} \int \frac{(8x + 4)dx}{4x^2 + 4x + 10} - \int \frac{dx}{\left(x + \frac{1}{2}\right)^2 + \left(\frac{3}{2}\right)^2}$$

$$\int \frac{6x - 1}{4x^2 + 4x + 10} dx = \frac{3}{4} ln(4x^2 + 4x + 10) - \frac{2}{3} tan^{-1}\left(\frac{x + \frac{1}{2}}{\frac{3}{2}}\right) + C$$

$$\int \frac{6x - 1}{4x^2 + 4x + 10} dx = \frac{3}{4} ln(4x^2 + 4x + 10) - \frac{2}{3} tan^{-1}\left(\frac{2x + 1}{3}\right) + C$$

$$\int \frac{6x-1}{4x^2+4x+10}dx = \frac{3}{4}ln[2(2x^2+2x+5)] - \frac{2}{3}tan^{-1}\left(\frac{2x+1}{3}\right) + C$$

$$\boxed{\int \frac{6x-1}{4x^2+4x+10}dx = \frac{3}{4}ln(2x^2+2x+5) - \frac{2}{3}tan^{-1}\left(\frac{2}{3}x+\frac{1}{3}\right) + C}$$

182

$$\int \frac{4x-3}{\sqrt{11+10x-x^2}}dx$$

The integral is rewritten   ***Equation 1***

$$\int \frac{4x-3}{\sqrt{11+10x-x^2}}dx = 4\int \frac{x\,dx}{\sqrt{11+10x-x^2}} - 3\int \frac{dx}{\sqrt{11+10x-x^2}}$$

The first integral is solved

$$4\int \frac{x\,dx}{\sqrt{11+10x-x^2}} \qquad u = 11+10x-x^2 \longrightarrow du = (10-2x)dx$$

The integral is multiplied by minus (-), multiplied and divided by 2 and added and subtracted 10.

$$-\frac{4}{2}\int \frac{-2x+10-10}{\sqrt{11+10x-x^2}}dx = -2\int \frac{(-2x+10)dx}{\sqrt{11+10x-x^2}} + 20\int \frac{dx}{\sqrt{11+10x-x^2}}$$

This result is substituted in equation 1 and is simplified

Ecuación 2

$$\int \frac{4x-3}{\sqrt{11+10x-x^2}}dx = -2\int \frac{(-2x+10)dx}{\sqrt{11+10x-x^2}} + 17\int \frac{dx}{\sqrt{11+10x-x^2}}$$

You fill squares in the second integral, and then you get the final result.

$$11+10x-x^2 = -[x^2-10x-11] = [6^2-(x-5)^2]$$

$$\int \frac{4x-3}{\sqrt{11+10x-x^2}}dx = -2\int \frac{(-2x+10)dx}{\sqrt{11+10x-x^2}} + 17\int \frac{dx}{\sqrt{[6^2-(x-5)^2]}}$$

$$\int \frac{4x-3}{\sqrt{11+10x-x^2}}\,dx = -2\int \frac{du}{\sqrt{u}} + 17\int \frac{dz}{\sqrt{a^2-z^2}}$$

$$\boxed{\int \frac{4x-3}{\sqrt{11+10x-x^2}}\,dx = -4\sqrt{-x^2+10x+11} + 17sen^{-1}\left(\frac{1}{6}x-\frac{5}{6}\right)+C}$$

$$\boxed{\int \sqrt{x^2+2x-3}\,dx} \qquad 183$$

It completes squares

$$x^2+2x-3 = (x^2+2x+1)-3-1 = (x+1)^2-4 = (x+1)^2-2^2$$

$$\int \sqrt{x^2+2x-3}\,dx = \int \sqrt{(x+1)^2-2^2}\,dx = \int \sqrt{u^2-a^2}\,du$$

$$\int \sqrt{x^2+2x-3}\,dx = \frac{x+1}{2}\sqrt{(x+1)^2-2^2} - \frac{2^2}{2}ln\left|(x+1)+\sqrt{(x+1)^2-2^2}\right|+C$$

$$\boxed{\int \sqrt{x^2+2x-3}\,dx = \frac{x+1}{2}\sqrt{x^2+2x-3} - 2ln\left|(x+1)+\sqrt{x^2+2x-3}\right|+C}$$

$$\boxed{\int \sqrt{x^2+4x}\,dx} \qquad 184$$

It completes squares

$$x^2+4x = x^2+4x+4-4 = (x+2)^2-2^2$$

$$\int \sqrt{x^2+4x}\,dx = \int \sqrt{(x+2)^2-2^2}\,dx$$

$$u = x+2 \ , \quad du = dx \ , \quad a = 2$$

$$\int \sqrt{x^2 + 4x} \; dx = \int \sqrt{u^2 - a^2} \; du$$

$$\int \sqrt{x^2 + 4x} \; dx = \frac{x+2}{2}\sqrt{(x+2)^2 - 2^2} - \frac{2^2}{2} \ln \left|(x+2) + \sqrt{(x+2)^2 - 2^2}\right| + C$$

$$\boxed{\int \sqrt{x^2 + 4x} \; dx = \frac{x+2}{2}\sqrt{x^2 + 4x} - 2\ln\left|(x+2) + \sqrt{x^2 + 4x}\right| + C}$$

---

185 $\qquad\qquad\qquad\qquad \boxed{\displaystyle\int \sqrt{12 + 4x - x^2} \; dx}$

It completes squares

$$12 + 4x - x^2 = -(x^2 - 4x - 12) = -[(x^2 - 4x + 4) - 12 - 4]$$

$$12 + 4x - x^2 = -[(x-2)^2 - 16] = 4^2 - (x-2)^2$$

$$\int \sqrt{12 + 4x - x^2} \; dx = \int \sqrt{4^2 - (x-2)^2} \; dx = \int \sqrt{a^2 - u^2} \; du$$

$$\int \sqrt{12 + 4x - x^2} \; dx = \frac{x-2}{2}\sqrt{4^2 - (x-2)^2} + \frac{4^2}{2} sen^{-1}\frac{(x-2)}{4} + C$$

$$\boxed{\int \sqrt{12 + 4x - x^2} \; dx = \frac{x-2}{2}\sqrt{12 + 4x - x^2} + 8sen^{-1}\left(\frac{1}{4}x - \frac{1}{2}\right) + C}$$

---

186 $\qquad\qquad\qquad\qquad\qquad \boxed{\displaystyle\int \sqrt{x^2 - 8x} \; dx}$

It completes squares

$$(x^2 - 8x + 16) - 16 = (x-4)^2 - 4^2$$

$$\int \sqrt{x^2 - 8x} \; dx = \int \sqrt{(x-4)^2 - 4^2} \; dx = \int \sqrt{u^2 - a^2} \; du$$

$$\int \sqrt{x^2 - 8x}\ dx = \frac{(x-4)}{2}\sqrt{(x-4)^2 - 16} - \frac{16}{2}\ln\left|(x-4) + \sqrt{(x-4)^2 - 16}\right| + C$$

$$\int \sqrt{x^2 - 8x}\ dx = \frac{(x-4)}{2}\sqrt{x^2 - 8x} - 8\ln\left|(x-4) + \sqrt{x^2 - 8x}\right| + C$$

$$\boxed{\int \sqrt{x^2 - 8x}\ dx = \frac{(x-4)}{2}\sqrt{x^2 - 8x} - 8\ln\left|(x-4) + \sqrt{x^2 - 8x}\right| + C}$$

$$\boxed{\int \frac{(5 - 4x)dx}{\sqrt{12x - 4x^2 - 8}}} \qquad 187$$

$$u = 12x - 4x^2 - 8 \longrightarrow du = (12 - 8x)dx$$

There are two ways to solve the integral one. The first is to multiply and divide the integral by two and then add two and subtract two. To find the du. The second way is to separate the integral into two integrals and solve. The authors chose the second form. It is suggested to the reader to apply the first form and compare the results and determine which of the forms is the most convenient.

Equation 1

$$\int \frac{(5 - 4x)dx}{\sqrt{12x - 4x^2 - 8}} = 5\int \frac{dx}{\sqrt{12x - 4x^2 - 8}} - 4\int \frac{x\ dx}{\sqrt{12x - 4x^2 - 8}}$$

The first integral is solved

$$5\int \frac{dx}{\sqrt{12x - 4x^2 - 8}}$$

It completes squares

$$12x - 4x^2 - 8 = -4(x^2 - 3x + 2)$$

$$= -4\left[(x^2 - 3x + \frac{9}{4}) + 2 - \frac{9}{4}\right]$$

$$= -4\left[(x - \frac{3}{2})^2 - \frac{1}{4}\right]$$

243

$$= -4\left[(x - \frac{3}{2})^2 - (\frac{1}{2})^2\right]$$

$$12x - 4x^2 - 8 = 4\left[\left(\frac{1}{2}\right)^2 - \left(x - \frac{3}{2}\right)^2\right]$$

$$5\int \frac{dx}{\sqrt{12x - 4x^2 - 8}} = 5\int \frac{dx}{\sqrt{4\left[\left(\frac{1}{2}\right)^2 - \left(x - \frac{3}{2}\right)^2\right]}} = \int \frac{du}{\sqrt{a^2 - u^2}}$$

$$5\int \frac{dx}{\sqrt{12x - 4x^2 - 8}} = \frac{5}{2} sen^{-1}\left(\frac{x - \frac{3}{2}}{\frac{1}{2}}\right) + C_1$$

$$\boxed{5\int \frac{dx}{\sqrt{12x - 4x^2 - 8}} = \frac{5}{2} sen^{-1}(2x - 3) + C_1 \quad \textbf{Equation 2}}$$

Solve the second integral of equation 1

$$4\int \frac{x\, dx}{\sqrt{12x - 4x^2 - 8}}$$

$$u = 12x - 4x^2 - 8 \qquad du = (12 - 8x)dx$$

$$4\int \frac{x\, dx}{\sqrt{12x - 4x^2 - 8}} = -\frac{4}{8}\int \frac{(-8x + 12 - 12)\, dx}{\sqrt{12x - 4x^2 - 8}}$$

$$4\int \frac{x\, dx}{\sqrt{12x - 4x^2 - 8}} = -\frac{1}{2}\int \frac{(-8x + 12)\, dx}{\sqrt{12x - 4x^2 - 8}} + 6\int \frac{dx}{\sqrt{12x - 4x^2 - 8}}$$

$$4\int \frac{x\, dx}{\sqrt{12x - 4x^2 - 8}} = -\frac{1}{2}\int \frac{du}{\sqrt{u}} + 3\int \frac{dx}{\sqrt{\left[\left(\frac{1}{2}\right)^2 - \left(x - \frac{3}{2}\right)^2\right]}}$$

$$4\int \frac{x\,dx}{\sqrt{12x-4x^2-8}} = -\sqrt{12x-4x^2-8} + 3sen^{-1}(2x-3) + C_2$$

Equation 3

Substitute the results of equation 2 and equation 3 in equation 1

$$\int \frac{(5-4x)dx}{\sqrt{12x-4x^2-8}} = \frac{5}{2}sen^{-1}(2x-3) + \sqrt{12x-4x^2-8} - 3sen^{-1}(2x-3) + C$$

$$\boxed{\int \frac{(5-4x)dx}{\sqrt{12x-4x^2-8}} = -\frac{1}{2}sen^{-1}(2x-3) + 2\sqrt{-x^2+3x-2} + C}$$

$$\boxed{\int \sqrt{6x - x^2}\,dx}$$

188

It completes squares

$$6x - x^2 = -x^2 + 6x = -(x^2 - 6x)$$

$$6x - x^2 = -[(x^2 - 6x + 9) - 9]$$

$$6x - x^2 = -[(x-3)^2 - 3^2]$$

$$6x - x^2 = 3^2 - (x-3)^2$$

$$\int \sqrt{6x - x^2}\,dx = \int \sqrt{3^2 - (x-3)^2}\,dx = \int \sqrt{a^2 - u^2}\,du$$

$$\int \sqrt{6x - x^2}\,dx = \frac{x-3}{2}\sqrt{9 - (x-3)^2} + \frac{9}{2}sen^{-1}\left(\frac{x-3}{3}\right) + C$$

$$\boxed{\int \sqrt{6x - x^2}\,dx = \frac{x-3}{2}\sqrt{6x - x^2} + \frac{9}{2}sen^{-1}\left(\frac{1}{3}x - 1\right) + C}$$

189

$$\int \frac{x \, dx}{\sqrt{27 + 6x - x^2}}$$

$$u = 27 + 6x - x^2 \ \longrightarrow \ du = (6 - 2x)dx$$

Multiply and divide by -2 the integral. The integral is added and subtracted 6.

$$\int \frac{x \, dx}{\sqrt{27 + 6x - x^2}} = -\frac{1}{2} \int \frac{(6 - 2x - 6) \, dx}{\sqrt{27 + 6x - x^2}} \qquad \textbf{\textit{Equation 1}}$$

$$\int \frac{x \, dx}{\sqrt{27 + 6x - x^2}} = -\frac{1}{2} \int \frac{(6 - 2x) \, dx}{\sqrt{27 + 6x - x^2}} + 3 \int \frac{dx}{\sqrt{27 + 6x - x^2}}$$

The resolution of the first integral is immediate. Now the second integral is working.

$$3 \int \frac{dx}{\sqrt{27 + 6x - x^2}}$$

It completes squares

$$27 + 6x - x^2 = -(x^2 - 6x - 27)$$

$$= -[(x^2 - 6x + 9) - 27 - 9]$$

$$= -[(x - 3)^2 - 36]$$

$$27 + 6x - x^2 = [36 - (x - 3)^2]$$

$$3 \int \frac{dx}{\sqrt{27 + 6x - x^2}} = 3 \int \frac{dx}{\sqrt{\left[6^2 - (x - 3)^2\right]}}$$

The previous result is substituted in equation 1, and the final result is obtained.

$$\int \frac{x \, dx}{\sqrt{27 + 6x - x^2}} = -\frac{1}{2} \int \frac{(6 - 2x) \, dx}{\sqrt{27 + 6x - x^2}} + 3 \int \frac{dx}{\sqrt{\left[6^2 - (x - 3)^2\right]}}$$

$$\int \frac{x\,dx}{\sqrt{27+6x-x^2}} = -\frac{1}{2}\int \frac{du}{\sqrt{u}} + 3\int \frac{dx}{\sqrt{\left[6^2-(x-3)^2\right]}}$$

$$\int \frac{x\,dx}{\sqrt{27+6x-x^2}} = -\frac{1}{2}\int u^{-\frac{1}{2}}du + 3\int \frac{dx}{\sqrt{\left[6^2-(x-3)^2\right]}}$$

$$\int \frac{x\,dx}{\sqrt{27+6x-x^2}} = -\sqrt{27+6x-x^2} + 3sen^{-1}\left(\frac{x-3}{6}\right) + C$$

$$\int \frac{x\,dx}{\sqrt{27+6x-x^2}} = -\sqrt{27+6x-x^2} + 3sen^{-1}\left(\frac{1}{6}x-\frac{1}{2}\right) + C$$

$$\boxed{\int \frac{x\,dx}{\sqrt{27+6x-x^2}} = -\sqrt{27+6x-x^2} + 3sen^{-1}\left(\frac{1}{6}x-\frac{1}{2}\right) + C}$$

$$\boxed{\int \frac{(x-1)\,dx}{3x^2-4x+3}}$$

190

$$u = 3x^2 - 4x + 3 \;\longrightarrow\; du = (6x-4)dx$$

Multiply and divide by 6 the integral

$$\int \frac{(x-1)\,dx}{3x^2-4x+3} = \frac{1}{6}\int \frac{(6x-6)\,dx}{3x^2-4x+3}$$

$$\int \frac{(x-1)\,dx}{3x^2-4x+3} = \frac{1}{6}\int \frac{(6x-4-2)\,dx}{3x^2-4x+3}$$

Then proceed to separate the integral into two integrals

Equation 1

$$\int \frac{(x-1)\,dx}{3x^2-4x+3} = \frac{1}{6}\int \frac{(6x-4)\,dx}{3x^2-4x+3} - \frac{1}{3}\int \frac{dx}{3x^2-4x+3}$$

The first integral is immediate. The second integral is solved

$$\frac{1}{3} \int \frac{dx}{3x^2 - 4x + 3}$$

It completes squares

$$3x^2 - 4x + 3 = 3\left(x^2 - \frac{4}{3}x + 1\right)$$

$$= 3\left(x^2 - \frac{4}{3}x + \frac{4}{9} + 1 - \frac{4}{9}\right)$$

$$= 3\left[\left(x - \frac{2}{3}\right)^2 + \frac{5}{9}\right]$$

$$3x^2 - 4x + 3 = 3\left[\left(x - \frac{2}{3}\right)^2 + \left(\frac{\sqrt{5}}{3}\right)^2\right]$$

$$\frac{1}{3} \int \frac{dx}{3x^2 - 4x + 3} = \frac{1}{9} \int \frac{dx}{\left[\left(x - \frac{2}{3}\right)^2 + \left(\frac{\sqrt{5}}{3}\right)^2\right]}$$

$$\frac{1}{3} \int \frac{dx}{3x^2 - 4x + 3} = \frac{1}{9} \frac{1}{\frac{\sqrt{5}}{3}} arctan \frac{x - \frac{2}{3}}{\frac{\sqrt{5}}{3}} + C$$

$$= \frac{3}{9\sqrt{5}} arctan \frac{3x - 2}{\sqrt{5}} + C$$

$$\frac{1}{3} \int \frac{d}{3x^2 - 4x + 3} = \frac{1}{3\sqrt{5}} arctan \frac{3x - 2}{\sqrt{5}} + C_2 \quad \textbf{\textit{Equation 2}}$$

Equation 2 is substituted in equation 1, and the final result is obtained.

$$\int \frac{(x - 1)\, dx}{3x^2 - 4x + 3} = \frac{1}{6} \int \frac{du}{u} - \frac{1}{3\sqrt{5}} arctan \frac{3x - 2}{\sqrt{5}} + C_2$$

$$\int \frac{(x-1)\,dx}{3x^2 - 4x + 3} = \frac{1}{6} ln[u] - \frac{1}{3\sqrt{5}} arctan \frac{3x - 2}{\sqrt{5}} + C$$

$$\int \frac{(x-1)\,dx}{3x^2 - 4x + 3} = \frac{1}{6} ln[3x^2 - 4x + 3] - \frac{1}{15}\sqrt{5}\, arctan\left(\frac{1}{5}(3x - 2)\sqrt{5}\right) + C$$

$$\boxed{\int \frac{(x-1)\,dx}{3x^2 - 4x + 3} = \frac{1}{6} ln[3x^2 - 4x + 3] - \frac{1}{15}\sqrt{5}\, arctan\left(\frac{1}{5}(3x - 2)\sqrt{5}\right) + C}$$

$$\boxed{\int \frac{(2x-3)\,dx}{x^2 + 6x + 15}}$$

$$u = x^2 + 6x + 15 \longrightarrow du = (2x + 6)dx$$

9 is added and 9 is subtracted

$$\int \frac{(2x-3)\,dx}{x^2 + 6x + 15} = \int \frac{(2x - 3 + 9 - 9)\,dx}{x^2 + 6x + 15}$$

$$\int \frac{(2x-3)\,dx}{x^2 + 6x + 15} = \int \frac{(2x + 6)\,dx}{x^2 + 6x + 15} - 9\int \frac{dx}{x^2 + 6x + 15} \qquad \textbf{\textit{Equation 1}}$$

The first integral is immediate. The second integral is worked It completes squares

$$9\int \frac{dx}{x^2 + 6x + 15} \quad \rightarrow \quad \textit{It completes squares}$$

$$x^2 + 6x + 15 = x^2 + 6x + 9 + 15 - 9 = (x^2 + 6x + 9) + 6$$

$$x^2 + 6x + 15 = (x + 3)^2 + (\sqrt{6})^2$$

$$9\int \frac{dx}{x^2 + 6x + 15} = 9\int \frac{dx}{(x + 3)^2 + (\sqrt{6})^2}$$

This partial result is substituted in equation 1 to obtain the final result.

$$\int \frac{(2x-3)\,dx}{x^2 + 6x + 15} = \int \frac{(2x + 6)\,dx}{x^2 + 6x + 15} - 9\int \frac{dx}{(x + 3)^2 + \left(\sqrt{6}\right)^2}$$

$$\int \frac{(2x-3)\,dx}{x^2+6x+15} = \int \frac{du}{u} - 9\int \frac{dz}{z^2+a^2}$$

$$\boxed{\int \frac{(2x-3)\,dx}{x^2+6x+15} = ln|x^2+6x+15| - \frac{3}{2}\sqrt{6}\,arctan\left(\frac{1}{6}(x+3)\sqrt{6}\right)+C}$$

192

$$\boxed{\int \frac{dx}{4x^2+4x+10}}$$

It completes squares

$$4x^2+4x+10 = 4\left(x^2+x+\frac{10}{4}\right)$$

$$4x^2+4x+10 = 4\left(x^2+x+\frac{1}{4}+\frac{10}{4}-\frac{1}{4}\right)$$

$$4x^2+4x+10 = 4\left[\left(x+\frac{1}{2}\right)^2+\frac{9}{4}\right]$$

$$4x^2+4x+10 = 4\left[\left(x+\frac{1}{2}\right)^2+\left(\frac{3}{2}\right)^2\right]$$

$$\int \frac{dx}{4x^2+4x+10} = \frac{1}{4}\int \frac{dx}{\left(x+\frac{1}{2}\right)^2+\left(\frac{3}{2}\right)^2}$$

$$\int \frac{dx}{4x^2+4x+10} = \frac{1}{4}\frac{1}{\frac{3}{2}}\,arctan\frac{x+\frac{1}{2}}{\frac{3}{2}}+C$$

$$\int \frac{dx}{4x^2+4x+10} = \frac{2}{12}\,arctan\frac{\frac{2x+1}{2}}{\frac{3}{2}}+C$$

$$\int \frac{dx}{4x^2 + 4x + 10} = \frac{1}{6} \ arctan \frac{2x + 1}{3} + C$$

$$\boxed{\int \frac{dx}{4x^2 + 4x + 10} = \frac{1}{6} \ arctan \left(\frac{2}{3}x + \frac{1}{3}\right) + C}$$

$$\boxed{\int \frac{(2x + 2) \ dx}{x^2 - 4x + 9}}$$

193

$$u = x^2 - 4x + 9 \quad \longrightarrow \quad du = (2x - 4)dx$$

$$\int \frac{(2x + 2) \ dx}{x^2 - 4x + 9} = \int \frac{(2x + 2 - 6 + 6) \ dx}{x^2 - 4x + 9}$$

$$= \int \frac{(2x + 2 - 6) \ dx}{x^2 - 4x + 9} + 6 \int \frac{dx}{x^2 - 4x + 9}$$

$$\int \frac{(2x + 2) \ dx}{x^2 - 4x + 9} = \int \frac{du}{u} + 6 \int \frac{dx}{x^2 - 4x + 9}$$

Fill squares in the second integral to obtain the final result.

$$x^2 - 4x + 9 = (x^2 - 4x + 4) + 9 - 4$$

$$x^2 - 4x + 9 = (x - 2)^2 + 5$$

$$x^2 - 4x + 9 = (x - 2)^2 + \left(\sqrt{5}\right)^2$$

$$\int \frac{(2x + 2) \ dx}{x^2 - 4x + 9} = \int \frac{du}{u} + 6 \int \frac{dx}{(x - 2)^2 + \left(\sqrt{5}\right)^2}$$

$$\int \frac{(2x + 2) \ dx}{x^2 - 4x + 9} = ln|u| + \frac{6}{\sqrt{5}} \ arctan \left(\frac{x - 2}{\sqrt{5}}\right) + C$$

$$\boxed{\int \frac{(2x + 2) \ dx}{x^2 - 4x + 9} = ln|x^2 - 4x + 9| + \frac{6}{5}\sqrt{5} \ arctan \left[\left(\frac{1}{5}x - \frac{2}{5}\right)\sqrt{5}\right] + C}$$

194

$$\int \frac{(2x + 4)\, dx}{\sqrt{4x - x^2}}$$

$$u = 4x - x^2 \ \longrightarrow \ du = (4 - 2x)dx$$

Multiply by minus (-) the integral, add and subtract 8

$$\int \frac{(2x + 4)\, dx}{\sqrt{4x - x^2}} = -\int \frac{(-2x - 4 + 8 - 8)\, dx}{\sqrt{4x - x^2}}$$

$$\int \frac{(2x + 4)\, dx}{\sqrt{4x - x^2}} = -\int \frac{(4 - 2x)dx}{\sqrt{4x - x^2}} + 8 \int \frac{dx}{\sqrt{4x - x^2}}$$

It completes squares in the second integral

$$4x - x^2 = -(x^2 - 4x) = -(x^2 - 4x + 4 - 4)$$

$$4x - x^2 = -[(x - 2)^2 - 4] = 2^2 - (x - 2)^2$$

$$\int \frac{(2x + 4)\, dx}{\sqrt{4x - x^2}} = -\int \frac{du}{u} + 8 \int \frac{dx}{\sqrt{2^2 - (x - 2)^2}}$$

$$\int \frac{(2x + 4)\, dx}{\sqrt{4x - x^2}} = -\int u^{-\frac{1}{2}}du + 8 \int \frac{dx}{\sqrt{2^2 - (x - 2)^2}}$$

$$\int \frac{(2x + 4)\, dx}{\sqrt{4x - x^2}} = -2u^{\frac{1}{2}} + 8\, arcsen\left(\frac{x - 2}{2}\right) + C$$

$$\boxed{\int \frac{(2x + 4)\, dx}{\sqrt{4x - x^2}} = -2\sqrt{4x - x^2} + 8\, arcsen\left(\frac{1}{2}x - 1\right) + C}$$

$$\frac{2}{3} \int \frac{\left(x + \frac{3}{2}\right) dx}{9x^2 - 12x + 8}$$

195

The integral problem is separated into two integrals

$$\frac{2}{3} \int \frac{\left(x + \frac{3}{2}\right) dx}{9x^2 - 12x + 8} = \frac{2}{3} \int \frac{x\, dx}{9x^2 - 12x + 8} + \frac{2}{3} * \frac{3}{2} \int \frac{dx}{9x^2 - 12x + 8}$$

Equation 1

$$\frac{2}{3} \int \frac{\left(x + \frac{3}{2}\right) dx}{9x^2 - 12x + 8} = \frac{2}{3} \int \frac{x\, dx}{9x^2 - 12x + 8} + \int \frac{dx}{9x^2 - 12x + 8}$$

The first integral is worked.

$$\frac{2}{3} \int \frac{x\, dx}{9x^2 - 12x + 8}$$

$$u = 9x^2 - 12x + 8 \;\longrightarrow\; du = (18x - 12)dx$$

Multiply and divide by 18 the integral, and add and subtract 12

$$\frac{2}{3} \int \frac{x\, dx}{9x^2 - 12x + 8} = \frac{2}{54} \int \frac{(18x - 12 + 12)\, dx}{9x^2 - 12x + 8}$$

$$\frac{2}{3} \int \frac{x\, dx}{9x^2 - 12x + 8} = \frac{1}{27} \int \frac{(18x - 12)\, dx}{9x^2 - 12x + 8} + \frac{12}{27} \int \frac{dx}{9x^2 - 12x + 8}$$

The partial result is replaced in equation 1

$$\frac{2}{3} \int \frac{\left(x + \frac{3}{2}\right) dx}{9x^2 - 12x + 8} = \frac{1}{27} \int \frac{(18x - 12)\, dx}{9x^2 - 12x + 8} + \frac{4}{9} \int \frac{dx}{9x^2 - 12x + 8} + \int \frac{dx}{9x^2 - 12x + 8}$$

$$\frac{2}{3} \int \frac{\left(x + \frac{3}{2}\right) dx}{9x^2 - 12x + 8} = \frac{1}{27} \int \frac{(18x - 12)\, dx}{9x^2 - 12x + 8} + \frac{13}{9} \int \frac{dx}{9x^2 - 12x + 8}$$

It completes squares in the second integral

$$9x^2 - 12x + 8 = 9\left(x^2 - \frac{12}{9}x + \frac{8}{9}\right) = 9\left(x^2 - \frac{12}{9}x + \frac{4}{9} + \frac{8}{9} - \frac{4}{9}\right)$$

$$9x^2 - 12x + 8 = 9\left[\left(x - \frac{2}{3}\right)^2 + \frac{4}{9}\right] = 9\left[\left(x - \frac{2}{3}\right)^2 + \left(\frac{2}{3}\right)^2\right]$$

Consequently, you have

$$\frac{2}{3}\int \frac{\left(x + \frac{3}{2}\right) dx}{9x^2 - 12x + 8} = \frac{1}{27}\int \frac{(18x - 12)\, dx}{9x^2 - 12x + 8} + \frac{13}{81}\int \frac{dx}{\left(x - \frac{2}{3}\right)^2 + \left(\frac{2}{3}\right)^2}$$

$$\frac{2}{3}\int \frac{\left(x + \frac{3}{2}\right) dx}{9x^2 - 12x + 8} = \frac{1}{27}\int \frac{du}{u} + \frac{13}{81}\int \frac{dx}{\left(x - \frac{2}{3}\right)^2 + \left(\frac{2}{3}\right)^2}$$

$$\frac{2}{3}\int \frac{\left(x + \frac{3}{2}\right) dx}{9x^2 - 12x + 8} = \frac{1}{27}\,ln|u| + \frac{13}{81} * \frac{1}{\frac{2}{3}}\, arctan\left(\frac{x - \frac{2}{3}}{\frac{2}{3}}\right) + C$$

$$\frac{2}{3}\int \frac{\left(x + \frac{3}{2}\right) dx}{9x^2 - 12x + 8} = \frac{1}{27}\,ln|9x^2 - 12x + 8| + \frac{39}{162}\, arctan\left(\frac{\frac{3x - 2}{3}}{\frac{2}{3}}\right) + C$$

$$\boxed{\frac{2}{3}\int \frac{\left(x + \frac{3}{2}\right) dx}{9x^2 - 12x + 8} = \frac{1}{27}\,ln|9x^2 - 12x + 8| + \frac{13}{54}\, arctan\left(\frac{3}{2}x - 1\right) + C}$$

$$\int \frac{(x + 6)\, dx}{\sqrt{5 - 4x - x^2}}$$

196

$$u = 5 - 4x - x^2 \longrightarrow du = (-4 - 2x)dx$$

Multiply and divide by -2 the integral

$$\int \frac{(x + 6)\, dx}{\sqrt{5 - 4x - x^2}} = -\frac{1}{2} \int \frac{(-2x - 12)\, dx}{\sqrt{5 - 4x - x^2}}$$

$$\int \frac{(x + 6)\, dx}{\sqrt{5 - 4x - x^2}} = -\frac{1}{2} \int \frac{(-2x - 4 - 8)\, dx}{\sqrt{5 - 4x - x^2}}$$

$$= -\frac{1}{2} \int \frac{(-2x - 4)\, dx}{\sqrt{5 - 4x - x^2}} + 4 \int \frac{dx}{\sqrt{5 - 4x - x^2}}$$

It completes squares in the second integral

$$5 - 4x - x^2 = -(x^2 + 4x - 5) = -(x^2 + 4x + 4 - 5 - 4)$$

$$5 - 4x - x^2 = -[(x + 2)^2 - 9] = 3^2 - (x + 2)^2$$

$$\int \frac{(x + 6)\, dx}{\sqrt{5 - 4x - x^2}} = -\frac{1}{2} \int \frac{(-2x - 4)\, dx}{\sqrt{5 - 4x - x^2}} + 4 \int \frac{dx}{\sqrt{3^2 - (x + 2)^2}}$$

$$\int \frac{(x + 6)\, dx}{\sqrt{5 - 4x - x^2}} = -\frac{1}{2} \int \frac{du}{\sqrt{u}} + 4 \int \frac{dx}{\sqrt{3^2 - (x + 2)^2}}$$

$$\int \frac{(x + 6)\, dx}{\sqrt{5 - 4x - x^2}} = -\frac{1}{2} \int u^{-\frac{1}{2}} + 4 \int \frac{dx}{\sqrt{3^2 - (x + 2)^2}}$$

$$\boxed{\int \frac{(x + 6)\, dx}{\sqrt{5 - 4x - x^2}} = -\sqrt{5 - 4x - x^2} + 4arcsen\left(\frac{1}{3}x + \frac{2}{3}\right) + C}$$

197

$$\int \frac{dx}{2x^2 + 20x + 60}$$

The integral is rewritten by taking common factor 2.

$$\int \frac{dx}{2x^2 + 20x + 60} = \int \frac{dx}{2(x^2 + 10x + 30)}$$

$$\int \frac{dx}{2x^2 + 20x + 60} = \frac{1}{2}\int \frac{dx}{x^2 + 10x + 30}$$

It completes squares

$$x^2 + 10x + 30 = x^2 + 10x + 25 + 30 - 25$$

$$x^2 + 10x + 30 = (x + 5)^2 + 5$$

$$x^2 + 10x + 30 = (x + 5)^2 + \sqrt{5}$$

$$\int \frac{dx}{2x^2 + 20x + 60} = \frac{1}{2}\int \frac{dx}{(x + 5)^2 + (\sqrt{5})^2}$$

$$\int \frac{dx}{2x^2 + 20x + 60} = \frac{1}{2} * \frac{1}{\sqrt{5}} arctan\left(\frac{x + 5}{\sqrt{5}}\right) + C$$

$$\int \frac{dx}{2x^2 + 20x + 60} = \frac{1}{10}\sqrt{5}\ arctan\left[\left(\frac{1}{5}x + 1\right)\sqrt{5}\right] + C$$

$$\boxed{\int \frac{dx}{2x^2 + 20x + 60} = \frac{1}{10}\sqrt{5}\ arctan\left[\left(\frac{1}{5}x + 1\right)\sqrt{5}\right] + C}$$

$$\int \frac{3\,dx}{\sqrt{80 + 32x - 4x^2}}$$

198

It completes squares

$$80 + 32x - 4x^2 = -4(x^2 - 8x - 20)$$

$$80 + 32x - 4x^2 = -4[(x^2 - 8x + 16) - 20 - 16]$$

$$80 + 32x - 4x^2 = 4[6^2 - (x - 4)^2]$$

$$\int \frac{3\,dx}{\sqrt{80 + 32x - 4x^2}} = \int \frac{3\,dx}{\sqrt{4[6^2 - (x - 4)^2]}} = \frac{3}{2}\int \frac{dx}{\sqrt{6^2 - (x - 4)^2}}$$

$$\boxed{\int \frac{3\,dx}{\sqrt{80 + 32x - 4x^2}} = \frac{3}{2}\,arcsen\left(\frac{1}{6}x - \frac{2}{3}\right) + C}$$

$$\int \frac{5\,dx}{\sqrt{28 - 12x - x^2}}$$

199

It completes squares

$$28 - 12x - x^2 = -(x^2 + 12x - 28) = -[(x^2 + 12x + 36) - 28 - 36]$$

$$28 - 12x - x^2 = -[(x + 6)^2 - 64] = [64 - (x + 6)^2] = [8^2 - (x + 6)^2]$$

$$\int \frac{5\,dx}{\sqrt{28 - 12x - x^2}} = 5\int \frac{dx}{\sqrt{8^2 - (x + 6)^2}}$$

$$\int \frac{5\,dx}{\sqrt{28 - 12x - x^2}} = 5\,arcsen\left(\frac{x + 6}{8}\right) + C$$

$$\boxed{\int \frac{5\,dx}{\sqrt{28 - 12x - x^2}} = 5\,arcsen\left(\frac{1}{8}x + \frac{3}{4}\right) + C}$$

200

$$\int \frac{dx}{\sqrt{12x - 4x^2 - 8}}$$

It completes squares

$$12x - 4x^2 - 8 = -4(x^2 - 3x + 2)$$

$$12x - 4x^2 - 8 = -4\left[\left(x^2 - 3x + \frac{9}{4}\right) + 2 - \frac{9}{4}\right]$$

$$12x - 4x^2 - 8 = -4\left[\left(x - \frac{3}{2}\right)^2 - \frac{1}{4}\right]$$

$$12x - 4x^2 - 8 = 4\left[\left(\frac{1}{2}\right)^2 - \left(x - \frac{3}{2}\right)^2\right]$$

$$\int \frac{dx}{\sqrt{12x - 4x^2 - 8}} = \int \frac{dx}{\sqrt{4\left[\left(\frac{1}{2}\right)^2 - \left(x - \frac{3}{2}\right)^2\right]}}$$

$$\int \frac{dx}{\sqrt{12x - 4x^2 - 8}} = \frac{1}{2}arcsen\frac{x - \frac{3}{2}}{\frac{1}{2}} + C$$

$$\int \frac{dx}{\sqrt{12x - 4x^2 - 8}} = \frac{1}{2}arcsen\frac{\frac{2x - 3}{2}}{\frac{1}{2}} + C$$

$$\int \frac{dx}{\sqrt{12x - 4x^2 - 8}} = \frac{1}{2}arcsen\frac{4x - 6}{2} + C$$

$$\boxed{\int \frac{dx}{\sqrt{12x - 4x^2 - 8}} = \frac{1}{2}\ arcsen(2x - 3) + C}$$

$$\int \frac{dx}{x^2 - 2x + 5}$$

201

It completes squares

$$x^2 - 2x + 5 = (x^2 - 2x + 1) + 5 - 1$$
$$x^2 - 2x + 5 = (x - 1)^2 + 4$$
$$x^2 - 2x + 5 = (x - 1)^2 + 2^2$$
$$\int \frac{dx}{x^2 - 2x + 5} = \int \frac{dx}{(x - 1)^2 + 2^2}$$
$$\int \frac{dx}{x^2 - 2x + 5} = \frac{1}{2} \arctan\left(\frac{x - 1}{2}\right) + C$$
$$\int \frac{dx}{x^2 - 2x + 5} = \frac{1}{2} \arctan\left(\frac{1}{2}x - \frac{1}{2}\right) + C$$

$$\boxed{\int \frac{dx}{x^2 - 2x + 5} = \frac{1}{2} \arctan\left(\frac{1}{2}x - \frac{1}{2}\right) + C}$$

$$\int \frac{(x - 1)\, dx}{\sqrt{8 + 2x - x^2}}$$

202

$$u = 8 + 2x - x^2 \longrightarrow du = (2 - 2x)dx$$

Multiply and divide by -2 the integral

$$\int \frac{(x - 1)\, dx}{\sqrt{8 + 2x - x^2}} = -\frac{1}{2} \int \frac{(-2x + 2)\, dx}{\sqrt{8 + 2x - x^2}}$$

$$\int \frac{(x - 1)\, dx}{\sqrt{8 + 2x - x^2}} = -\frac{1}{2} \int \frac{du}{\sqrt{u}}$$

$$\int \frac{(x-1)\,dx}{\sqrt{8+2x-x^2}} = -\frac{1}{2}\int u^{-\frac{1}{2}}\,du$$

$$\int \frac{(x-1)\,dx}{\sqrt{8+2x-x^2}} = -\frac{1}{2}\frac{u^{\frac{1}{2}}}{\frac{1}{2}} + C$$

$$\int \frac{(x-1)\,dx}{\sqrt{8+2x-x^2}} = -u^{\frac{1}{2}} + C$$

$$\boxed{\int \frac{(x-1)\,dx}{\sqrt{8+2x-x^2}} = -\sqrt{8+2x-x^2} + C}$$

203 $$\int \sqrt{12 - 8x - 4x^2}\,dx$$

It completes squares

$$12 - 8x - 4x^2 = -4(x^2 + 2x - 3) = -4[(x^2 + 2x + 1) - 3 - 1]$$

$$12 - 8x - 4x^2 = -4[(x+1)^2 - 4] = -4[(x+1)^2 - 2^2]$$

$$12 - 8x - 4x^2 = 4\left[2^2 - (x+1)^2\right]$$

$$\int \sqrt{12 - 8x - 4x^2}\,dx = \int \sqrt{4\left[2^2 - (x+1)^2\right]}\,dx$$

$$\int \sqrt{12 - 8x - 4x^2}\,dx = 2\int \sqrt{2^2 - (x+1)^2}\,dx$$

$$\int \sqrt{12 - 8x - 4x^2}\,dx = \frac{x+1}{2}\sqrt{2^2 - (x+1)^2} + \frac{2^2}{2}sen^{-1}\left(\frac{x+1}{2}\right) + C$$

$$\int \sqrt{12 - 8x - 4x^2}\,dx = \frac{x+1}{2}\sqrt{-x^2 - 2x + 3} + 2sen^{-1}\left(\frac{x+1}{2}\right) + C$$

$$\int \sqrt{12 - 8x - 4x^2}\ dx = (x + 1)\sqrt{-x^2 - 2x + 3} + 4sen^{-1}\left(\frac{1}{2}x + \frac{1}{2}\right) + C$$

$$\int \sqrt{12 - 8x - 4x^2}\ dx = (x + 1)\sqrt{-x^2 - 2x + 3} + 4sen^{-1}\left(\frac{1}{2}x + \frac{1}{2}\right) + C$$

$$\int \sqrt{x^2 - x + \frac{5}{4}}\ dx$$

204

It completes squares

$$x^2 - x + \frac{5}{4} = x^2 - x + \frac{1}{4} + \frac{5}{4} - \frac{1}{4}$$

$$x^2 - x + \frac{5}{4} = \left(x - \frac{1}{2}\right)^2 + 1$$

$$x^2 - x + \frac{5}{4} = \left(x - \frac{1}{2}\right)^2 + 1^2$$

$$\int \sqrt{x^2 - x + \frac{5}{4}}\ dx = \int \sqrt{\left(x - \frac{1}{2}\right)^2 + 1^2}\ dx$$

$$\int \sqrt{x^2 - x + \frac{5}{4}}\ dx = \frac{x - \frac{1}{2}}{2}\sqrt{x^2 - x + \frac{5}{4}} + \frac{1}{2}\ln\left|\left(x - \frac{1}{2}\right) + \sqrt{x^2 - x + \frac{5}{4}}\right| + C$$

$$\int \sqrt{x^2 - x + \frac{5}{4}}\ dx = \frac{(2x - 1)}{4}\sqrt{x^2 - x + \frac{5}{4}} + \frac{1}{2}\ln\left|\left(x - \frac{1}{2}\right) + \sqrt{x^2 - x + \frac{5}{4}}\right| + C$$

205

$$\int \frac{x \, dx}{x^2 + 4x + 5}$$

$$u = x^2 + 4x + 5 \quad du = (2x + 4)dx$$

Multiply and divide by 2, add and subtract 4, all that in the integral

$$\int \frac{x \, dx}{x^2 + 4x + 5} = \frac{1}{2} \int \frac{(2x + 4 - 4) \, dx}{x^2 + 4x + 5}$$

$$\int \frac{x \, dx}{x^2 + 4x + 5} = \int \frac{(2x + 4) \, dx}{x^2 + 4x + 5} - 2 \int \frac{dx}{x^2 + 4x + 5}$$

$$\int \frac{x \, dx}{x^2 + 4x + 5} = \int \frac{du}{u} - 2 \int \frac{dx}{x^2 + 4x + 5} \qquad \textit{Equation 1}$$

The first integral is immediate. The second integral is worked

$$\int \frac{dx}{x^2 + 4x + 5}$$

It completes squares

$$x^2 + 4x + 5 = (x^2 + 4x + 4) + 5 - 4$$

$$x^2 + 4x + 5 = (x + 2)^2 + 1^2$$

This partial result is replaced in equation 1

$$\int \frac{x \, dx}{x^2 + 4x + 5} = \int \frac{du}{u} - 2 \int \frac{dx}{(x + 2)^2 + 1^2}$$

$$\boxed{\int \frac{x \, dx}{x^2 + 4x + 5} = \frac{1}{2} ln\left|x^2 + 4x + 5\right| - 2 \arctan(x + 2) + C}$$

$$\int \frac{(2x+3)\,dx}{4x^2+4x+5}$$

206

$$u = 4x^2 + 4x + 5 \qquad du = (8x+4)dx$$

Multiply and divide by 4 the integral

$$\int \frac{(2x+3)\,dx}{4x^2+4x+5} = \frac{1}{4}\int \frac{(8x+12)\,dx}{4x^2+4x+5} = \frac{1}{4}\int \frac{(8x+4+8)\,dx}{4x^2+4x+5}$$

$$\int \frac{(2x+3)\,dx}{4x^2+4x+5} = \frac{1}{4}\int \frac{(8x+4)\,dx}{4x^2+4x+5} + 2\int \frac{dx}{4x^2+4x+5}$$

$$\int \frac{(2x+3)\,dx}{4x^2+4x+5} = \frac{1}{4}\int \frac{du}{u} + 2\int \frac{dx}{4x^2+4x+5} \qquad \textbf{\textit{Equation} 1}$$

The first integral is immediate. The second integral is worked

$$\int \frac{dx}{4x^2+4x+5}$$

It completes squares

$$4x^2 + 4x + 5 = 4\left(x^2 + x + \frac{5}{4}\right) = 4\left[\left(x^2 + x + \frac{1}{4}\right) + \frac{5}{4} - \frac{1}{4}\right]$$

$$4x^2 + 4x + 5 = 4\left[\left(x + \frac{1}{2}\right)^2 + 1^2\right]$$

This partial result is replaced in equation 1

$$\int \frac{(2x+3)\,dx}{4x^2+4x+5} = \frac{1}{4}\int \frac{du}{u} + \frac{1}{2}\int \frac{dx}{\left(x+\frac{1}{2}\right)^2 + 1^2}$$

$$\boxed{\int \frac{(2x+3)\,dx}{4x^2+4x+5} = \frac{1}{4}ln|4x^2+4x+5| + \frac{1}{2}arctan\left(x+\frac{1}{2}\right) + C}$$

207

$$\int \frac{(x+2)\,dx}{x^2+2x+2}$$

$$u = x^2 + 2x + 2 \;\longrightarrow\; du = (2x+2)dx$$

Multiply and divide by 2 the integral

$$\int \frac{(x+2)\,dx}{x^2+2x+2} = \frac{1}{2}\int \frac{(2x+4)\,dx}{x^2+2x+2} = \frac{1}{2}\int \frac{(2x+2+2)\,dx}{x^2+2x+2}$$

$$\int \frac{(x+2)\,dx}{x^2+2x+2} = \frac{1}{2}\int \frac{(2x+2)\,dx}{x^2+2x+2} + \int \frac{dx}{x^2+2x+2}$$

$$\int \frac{(x+2)\,dx}{x^2+2x+2} = \frac{1}{2}\int \frac{du}{u} + \int \frac{dx}{x^2+2x+2} \qquad \textbf{\textit{Equation 1}}$$

The first integral is immediate. The second integral is worked

$$\int \frac{dx}{x^2+2x+2}$$

It completes squares

$$x^2 + 2x + 2 = (x^2 + 2x + 1) + 2 - 1 = (x+1)^2 + 1^2$$

This partial result is substituted in equation 1.

$$\int \frac{(x+2)\,dx}{x^2+2x+2} = \frac{1}{2}\int \frac{du}{u} + \int \frac{dx}{(x+1)^2 + 1^2}$$

$$\int \frac{(x+2)\,dx}{x^2+2x+2} = \frac{1}{2}\,ln|x^2+2x+2| + \arctan(x+1) + C$$

$$\boxed{\int \frac{(x+2)\,dx}{x^2+2x+2} = \frac{1}{2}\,ln\left|x^2+2x+2\right| + \arctan(x+1) + C}$$

$$\int \frac{(2x + 1)\, dx}{x^2 + 8x - 2}$$

208

$$u = x^2 + 8x - 2 \longrightarrow du = (2x + 8)dx$$

7 is added and subtracted in the integral

$$\int \frac{(2x + 1)\, dx}{x^2 + 8x - 2} = \int \frac{(2x + 1 + 7 - 7)\, dx}{x^2 + 8x - 2}$$

$$\int \frac{(2x + 1)\, dx}{x^2 + 8x - 2} = \int \frac{(2x + 8)\, dx}{x^2 + 8x - 2} - 7 \int \frac{dx}{x^2 + 8x - 2}$$

$$\int \frac{(2x + 1)\, dx}{x^2 + 8x - 2} = \int \frac{du}{u} - 7 \int \frac{dx}{x^2 + 8x - 2} \quad \textbf{\textit{Equation 1}}$$

The first integral is immediate. The second integral is worked

$$\int \frac{dx}{x^2 + 8x - 2}$$

It completes squares

$$x^2 + 8x - 2 = (x^2 + 8x + 16) - 2 - 16$$

$$x^2 + 8x - 2 = (x + 4)^2 - 18$$

$$x^2 + 8x - 2 = (x + 4)^2 - (\sqrt{18})^2$$

This partial result is replaced in equation 1

$$\int \frac{(2x + 1)\, dx}{x^2 + 8x - 2} = \int \frac{du}{u} - 7 \int \frac{dx}{x^2 + 8x - 2} \quad \textbf{\textit{Equation 1}}$$

$$\int \frac{(2x + 1)dx}{x^2 + 8x - 2} = \int \frac{du}{u} - 7 \int \frac{dx}{(x + 4)^2 - \left(\sqrt{18}\right)^2}$$

$$\int \frac{(2x + 1)\, dx}{x^2 + 8x - 2} = \int \frac{du}{u} - 7 \int \frac{dx}{u^2 - a^2}$$

$$\int \frac{(2x+1)\,dx}{x^2+8x-2} = ln(u) - 7\int \frac{dx}{u^2-a^2}$$

$$\int \frac{(2x+1)\,dx}{x^2+8x-2} = ln|x^2+8x-2| - 7\,\frac{1}{2\sqrt{18}}\,ln\left|\frac{(x+4)-\sqrt{18}}{(x+4)+\sqrt{18}}\right| + C$$

$$\int \frac{(2x+1)\,dx}{x^2+8x-2} = ln|x^2+8x-2| - \frac{7\sqrt{18}}{36}\,ln\left|\frac{(x+4)-\sqrt{18}}{(x+4)+\sqrt{18}}\right| + C$$

$$\int \frac{(2x+1)\,dx}{x^2+8x-2} = ln|x^2+8x-2| - \frac{7\sqrt{18}}{36}\,ln\left|\frac{(x+4)-3\sqrt{2}}{(x+4)+3\sqrt{2}}\right| + C$$

$$\boxed{\int \frac{(2x+1)\,dx}{x^2+8x-2} = ln|x^2+8x-2| - \frac{7\sqrt{2}}{12}\,ln\left|\frac{(x+4)-3\sqrt{2}}{(x+4)+3\sqrt{2}}\right| + C}$$

209

$$\boxed{\int \frac{(x-1)\,dx}{x^2+2x+2}}$$

The integral is separated into two integrals

$$\int \frac{(x-1)\,dx}{x^2+2x+2} = \int \frac{x\,dx}{x^2+2x+2} - \int \frac{dx}{x^2+2x+2} \qquad \textbf{\textit{Equation 1}}$$

The first integral of equation 1 is worked

$$\int \frac{x\,dx}{x^2+2x+2}$$

$$u = x^2+2x+2 \longrightarrow du = (2x+2)dx$$

Multiply and divide by 2, add and subtract 2 in the integral

$$\int \frac{x\,dx}{x^2+2x+2} = \frac{1}{2}\int \frac{(2x+2-2)dx}{x^2+2x+2}$$

$$\int \frac{x\,dx}{x^2+2x+2} = \frac{1}{2}\int \frac{(2x+2)\,dx}{x^2+2x+2} - \int \frac{dx}{x^2+2x+2}$$

$$\int \frac{x\,dx}{x^2 + 2x + 2} = \frac{1}{2}\int \frac{du}{u} - \int \frac{dx}{x^2 + 2x + 2}$$

The second integral of the previous partial result is worked

$$\int \frac{dx}{x^2 + 2x + 2}$$

It completes squares

$$x^2 + 2x + 2 = (x^2 + 2x + 1) + 2 - 1 = (x + 1)^2 + 1^2$$

$$\int \frac{dx}{x^2 + 2x + 2} = \int \frac{dx}{(x + 1)^2 + 1^2}$$

Consequently, you have

$$\int \frac{x\,dx}{x^2 + 2x + 2} = \frac{1}{2}\int \frac{du}{u} - \int \frac{dx}{(x + 1)^2 + 1^2}$$

$$\int \frac{x\,dx}{x^2 + 2x + 2} = \frac{1}{2}ln|x^2 + 2x + 2| - \arctan(x + 1) + C_1 \;\; \textbf{Equation 2}$$

The result of the second integral of equation 1 is

$$\int \frac{dx}{x^2 + 2x + 2} = \int \frac{dx}{(x + 1)^2 + 1^2} = \arctan(x + 1) + C_2 \;\; \textbf{Equation 3}$$

Replace the result of equation 2 and of equation 3 in equation 1

$$\int \frac{(x - 1)\,dx}{x^2 + 2x + 2} = \int \frac{x\,dx}{x^2 + 2x + 2} - \int \frac{dx}{x^2 + 2x + 2} \;\; \textbf{Equation 1}$$

$$\int \frac{(x - 1)\,dx}{x^2 + 2x + 2} = \frac{1}{2}ln|x^2 + 2x + 2| - \arctan(x + 1) + C_1 - \arctan(x + 1) + C_2$$

$$\int \frac{(x - 1)\,dx}{x^2 + 2x + 2} = \frac{1}{2}ln|x^2 + 2x + 2| - 2\arctan(x + 1) + C$$

$$\boxed{\int \frac{(x - 1)\,dx}{x^2 + 2x + 2} = \frac{1}{2}ln|x^2 + 2x + 2| - 2\arctan(x + 1) + C}$$

In the following exercises are the integrals of irrational functions

$$\int \frac{dx}{x - \sqrt[3]{x}}$$

210

The only irrational expression is $\sqrt[3]{x}$. Then the next variable change is made $x = z^n$,
Where n is the index of the root.

$$x = z^3 \quad , \quad dx = 3z^2 dz \quad , \quad \sqrt[3]{x} = z$$

Therefore, you have

$$\int \frac{dx}{x - \sqrt[3]{x}} = \int \frac{3z^2 dz}{z^3 - z} = 3\int \frac{z^2 dz}{z(z^2 - 1)}$$

$$\int \frac{dx}{x - \sqrt[3]{x}} = 3\int \frac{z \, dz}{z^2 - 1}$$

Where $t = z^2 - 1 \quad \to \quad dt = 2zdz$

$$\int \frac{dx}{x - \sqrt[3]{x}} = \frac{3}{2}\int \frac{dt}{t} = \frac{3}{2} ln|t| + C$$

$$\int \frac{dx}{x - \sqrt[3]{x}} = \frac{3}{2} ln|z^2 - 1| + C$$

To return to the original variable, we operate in the following way: The cube root is extracted to both members of the equation. $x = z^3 \ \to \ \sqrt[3]{x} = z$. Then it is replaced in the term z^2.

$$\boxed{\int \frac{dx}{x - \sqrt[3]{x}} = \frac{3}{2} ln \left| x^{\frac{2}{3}} - 1 \right| + C}$$

211

$$\int \frac{\sqrt[3]{x}+1}{\sqrt[3]{x}-1}\, dx$$

In the integral, two irrational expressions of the same index are presented. Then we make a change of variable: $x = z^n$. Where n is the root index.

Therefore, you have $x = z^3$, $dx = 3z^2 dz$, $\sqrt[3]{x} = z$

$$\int \frac{\sqrt[3]{x}+1}{\sqrt[3]{x}-1}\, dx = \int \frac{(z+1)3z^2 dz}{z-1}$$

$$\int \frac{\sqrt[3]{x}+1}{\sqrt[3]{x}-1}\, dx = \int \frac{(3z^3+3z^2)dz}{z-1} = 3\int \frac{z^3+z^2}{z-1}\, dz$$

To solve the resulting integral, we divide the polynomial of the numerator with the polynomial of the denominator. Resulting

$$\frac{z^3+z^2}{z-1} = z^2 + 2z + 2 + \frac{2}{z-1}$$

As a result, you get

$$3\int \frac{z^3+z^2}{z-1}\, dz = 3\int z^2 dz + 6\int z\, dz + 6\int dz + 6\int \frac{dz}{z-1}$$

$$3\int \frac{z^3+z^2}{z-1}\, dz = z^3 + 3z^2 + 6z + 6\ln|z-1| + C$$

$$\int \frac{\sqrt[3]{x}+1}{\sqrt[3]{x}-1}\, dx = (\sqrt[3]{x})^3 + 3(\sqrt[3]{x})^2 + 6\sqrt[3]{x} + 6\ln|\sqrt[3]{x}-1| + C$$

$$\boxed{\int \frac{\sqrt[3]{x}+1}{\sqrt[3]{x}-1}\, dx = x + 3x^{\frac{2}{3}} + 6x^{\frac{1}{3}} + 6\ln\left|x^{\frac{1}{3}}-1\right| + C}$$

$$\int \frac{dx}{\sqrt{x} + \sqrt[3]{x}}$$

212

In the integral two irrational expressions of different index are presented. That means we have to determine the least common multiple (m.c.m.) of the indexes (2 and 3) of the roots.

Therefore, the m.c.m is 6. Then we make a change of variable:: $x = z^n$. Where n is the least common multiple.

Therefore, you have $x = z^6$, $dx = 6z^5 dz$, $\sqrt[6]{x} = z$

$$\int \frac{dx}{\sqrt{x} + \sqrt[3]{x}} = \int \frac{dx}{\sqrt[6]{x^3} + \sqrt[6]{x^2}} = \int \frac{dx}{\sqrt[6]{z^{18}} + \sqrt[6]{z^{12}}}$$

$$\int \frac{dx}{\sqrt{x} + \sqrt[3]{x}} = \int \frac{6z^5 dz}{z^3 + z^2} = 6 \int \frac{z^5 dz}{z^2(z+1)} = 6 \int \frac{z^3 dz}{z+1}$$

$$\int \frac{dx}{\sqrt{x} + \sqrt[3]{x}} = 6 \int \left(z^2 - z + 1 - \frac{1}{z+1} \right) dz$$

$$\int \frac{dx}{\sqrt{x} + \sqrt[3]{x}} = 6 \int z^2 dz - 6 \int z dz + 6 \int dz - 6 \int \frac{dz}{z+1}$$

$$\int \frac{dx}{\sqrt{x} + \sqrt[3]{x}} = 2z^3 - 3z^2 + 6z - 6ln|z+1| + C$$

$$\int \frac{dx}{\sqrt{x} + \sqrt[3]{x}} = 2\sqrt{x} - 3\sqrt[3]{x} + 6\sqrt[6]{x} - 6ln|\sqrt[6]{x} + 1| + C$$

$$\boxed{\int \frac{dx}{\sqrt{x} + \sqrt[3]{x}} = 2x^{\frac{1}{2}} - 3x^{\frac{1}{3}} + 6x^{\frac{1}{6}} - 6ln\left|x^{\frac{1}{6}} + 1\right| + C}$$

213

$$\int \frac{dx}{\sqrt{x} + \sqrt[4]{x}}$$

In the integral two irrational expressions of different index are presented. That means we have to determine the least common multiple (m.c.m.) of the indexes (2 and 4) of the roots. Therefore, the m.c.m is 4. Then we make a change of variable: $x = z^n$. Where n is the least common multiple.

Therefore, you have $\qquad x = z^4$, $\qquad dx = 4z^3 dz$, $\qquad \sqrt[4]{x} = z$

$$\int \frac{dx}{\sqrt{x} + \sqrt[4]{x}} = \int \frac{dx}{\sqrt[4]{x^2} + \sqrt[4]{x}} = \int \frac{dx}{\sqrt[4]{z^8} + \sqrt[4]{z^4}}$$

$$\int \frac{dx}{\sqrt{x} + \sqrt[4]{x}} = \int \frac{4z^3 dz}{z^2 + z} = 4 \int \frac{z^3 dz}{z(z+1)}$$

$$\int \frac{dx}{\sqrt{x} + \sqrt[4]{x}} = 4 \int \frac{z^2 dz}{z+1}$$

$$\int \frac{dx}{\sqrt{x} + \sqrt[4]{x}} = 4 \int \left(z - 1 + \frac{1}{z+1} \right) dz$$

$$\int \frac{dx}{\sqrt{x} + \sqrt[4]{x}} = 4 \int z \, dz - 4 \int dz + 4 \int \frac{dz}{z+1}$$

$$\int \frac{dx}{\sqrt{x} + \sqrt[4]{x}} = 2z^2 - 4z + 4ln|z+1| + C$$

$$\boxed{\int \frac{dx}{\sqrt{x} + \sqrt[4]{x}} = 2\sqrt{x} - 4\sqrt[4]{x} + 4ln|\sqrt[4]{x} + 1| + C}$$

$$\int \frac{\sqrt{x}\, dx}{1 + \sqrt[3]{x}}$$

214

In the integral two irrational expressions of different index are presented. That means we have to determine the least common multiple (m.c.m.) of the indexes (2 and 3) of the roots. Therefore, the m.c.m is 6. Then we make a change of variable:$x = z^n$. Where n is the least common multiple.

Therefore, you have: $x = z^6$, $\qquad dx = 6z^5 dz$, $\qquad \sqrt[6]{x} = z$

$$\int \frac{\sqrt{x}\, dx}{1 + \sqrt[3]{x}} = \int \frac{\sqrt[6]{x^3}\, dx}{1 + \sqrt[6]{x^2}} = \int \frac{z^3 6z^5 dz}{1 + z^2} = 6 \int \frac{z^8 dz}{z^2 + 1}$$

$$\int \frac{\sqrt{x}\, dx}{1 + \sqrt[3]{x}} = 6 \int \left(z^6 - z^4 + z^2 - 1 + \frac{1}{z^2 + 1} \right)$$

$$\int \frac{\sqrt{x}\, dx}{1 + \sqrt[3]{x}} = 6 \int z^6 dz - 6 \int z^4 dz + 6 \int z^2 dz - 6 \int dz + 6 \int \frac{dz}{z^2 + 1}$$

$$\int \frac{\sqrt{x}\, dx}{1 + \sqrt[3]{x}} = \frac{6}{7}z^7 - \frac{6}{5}z^5 + 2z^3 - 6z + 6tan^{-1}(z) + C$$

$$\int \frac{\sqrt{x}\, dx}{1 + \sqrt[3]{x}} = \frac{6}{7}\left(\sqrt[6]{x}\right)^7 - \frac{6}{5}\left(\sqrt[6]{x}\right)^5 + 2\sqrt[6]{x} - 6\sqrt[6]{x} + 6tan^{-1}(z) + C$$

$$\boxed{\int \frac{\sqrt{x}\, dx}{1 + \sqrt[3]{x}} = \frac{6}{7}x^{\frac{7}{6}} - \frac{6}{5}x^{\frac{5}{6}} + 2x^{\frac{1}{6}} - 6x^{\frac{1}{6}} + 6tan^{-1}\left(x^{\frac{1}{6}}\right) + C}$$

215

$$\int \frac{x\,dx}{\sqrt[5]{3x+2}}$$

The change of variable is

$$3x + 2 = z^5 \;\rightarrow\; 3x = z^5 - 2 \;\rightarrow\; x = \frac{1}{3}z^5 - \frac{2}{3} \;\rightarrow\; dx = \frac{5}{3}z^4 dz$$

$$\int \frac{x\,dx}{\sqrt[5]{3x+2}} = \int \frac{\left(\frac{1}{3}z^5 - \frac{2}{3}\right)\frac{5}{3}z^4 dz}{z}$$

$$\int \frac{x\,dx}{\sqrt[5]{3x+2}} = \int \frac{\left(\frac{5}{9}z^9 - \frac{10}{9}z^4\right)dz}{z} = \frac{5}{9}\int \frac{z^9}{z}\,dz - \frac{10}{9}\int \frac{z^4}{z}\,dz$$

$$\int \frac{x\,dx}{\sqrt[5]{3x+2}} = \frac{5}{9}\int z^8\,dz - \frac{10}{9}\int z^3\,dz \;\rightarrow\; Each\ integral\ is\ solved$$

$$\int \frac{x\,dx}{\sqrt[5]{3x+2}} = \frac{5}{81}z^9 - \frac{10}{36}z^4 + C$$

Then, the respective substitutions are made to return to the original variable

$$\int \frac{x\,dx}{\sqrt[5]{3x+2}} = \frac{5}{81}\left(\sqrt[5]{3x+2}\right)^9 - \frac{5}{18}\left(\sqrt[5]{3x+2}\right)^4 + C$$

$$\boxed{\int \frac{x\,dx}{\sqrt[5]{3x+2}} = \frac{5}{81}(3x+2)^{\frac{9}{5}} - \frac{5}{18}(3x+2)^{\frac{4}{5}} + C}$$

$$\int x \sqrt[3]{x-4} \; dx$$

216

The change of variable is

$$x - 4 = z^3 \quad \rightarrow \quad x = z^3 + 4 \quad \rightarrow \quad dx = 3z^2 dz \;, z = \sqrt[3]{x-4}$$

$$\int x \sqrt[3]{x-4} \; dx = \int (z^3 + 4) \, z \; 3z^2 dz$$

$$\int x \sqrt[3]{x-4} \; dx = \int (3z^6 + 12z^3) dz$$

$$\int x \sqrt[3]{x-4} \; dx = 3 \int z^6 dz + 12 \int z^3 dz$$

$$\int x \sqrt[3]{x-4} \; dx = \frac{3}{7} z^7 + \frac{12}{4} z^4 + C$$

Then, the respective substitutions are made to return to the original variable

$$\int x \sqrt[3]{x-4} \; dx = \frac{3}{7} \left(\sqrt[3]{x-4}\right)^7 + 3\left(\sqrt[3]{x-4}\right)^4 + C$$

$$\int x \sqrt[3]{x-4} \; dx = \frac{3}{7} (x-4)^{\frac{7}{3}} + 3(x-4)^{\frac{4}{3}} + C$$

$$\int x \sqrt[3]{x-4} \; dx = \frac{3}{7} (x-4)^{\frac{4}{3}} [x - 4 + 7] + C$$

$$\boxed{\int x \sqrt[3]{x-4} \; dx = \frac{3}{7} (x-4)^{\frac{4}{3}} [x + 3] + C}$$

217

$$\int \frac{dx}{x(\sqrt{x-1}-1)}$$

The change of variable is

$$x - 1 = z^2 \quad \rightarrow \quad x = z^2 + 1 \quad \rightarrow \quad dx = 2z\,dz \quad \rightarrow \quad z = \sqrt{x-1}$$

$$\int \frac{dx}{x(\sqrt{x-1}-1)} = 2\int \frac{z\,dz}{(z^2+1)(z-1)}$$

The resulting integral is solved by applying the partial fraction technique.

$$\frac{z\,dz}{(z^2+1)(z-1)} = \frac{Az+B}{z^2+1} + \frac{C}{z-1}$$

$$z = (Az+B)(z-1) + C(z^2+1)$$

$$z = Az^2 - Az + Bz - B + Cz^2 + C$$

$$z = (A+C)z^2 + (-A+B)z + (-B+C)$$

From the above, the following system of equations originates.

$$
\begin{aligned}
A & & + C & = 0 \\
-A & + B & & = 1 \\
& -B & + C & = 0
\end{aligned}
$$

The system of equations is solved and the following values are obtained.

$$A = -\frac{1}{2} \quad , \quad B = \frac{1}{2} \quad y \quad C = \frac{1}{2}$$

$$\int \frac{z\,dz}{(z^2+1)(z-1)} = \int \frac{\left(-\frac{1}{2}z+\frac{1}{2}\right)dz}{z^2+1} + \int \frac{\frac{1}{2}dz}{z-1}$$

$$\int \frac{z\,dz}{(z^2+1)(z-1)} = -\frac{1}{2}\int \frac{z\,dz}{z^2+1} + \frac{1}{2}\int \frac{dz}{z^2+1} + \frac{1}{2}\int \frac{dz}{z-1}$$

$$\int \frac{z\,dz}{(z^2+1)(z-1)} = -\frac{1}{4}\int \frac{du}{u} + \frac{1}{2}\int \frac{dz}{z^2+1} + \frac{1}{2}\int \frac{dt}{t}$$

$$\int \frac{z\,dz}{(z^2+1)(z-1)} = -\frac{1}{4}ln|u| + \frac{1}{2}tan^{-1}z + \frac{1}{2}ln|t| + C$$

$$\int \frac{z\,dz}{(z^2+1)(z-1)} = -\frac{1}{4}ln|z^2+1| + \frac{1}{2}tan^{-1}z + \frac{1}{2}ln|z-1| + C$$

$$\int \frac{dx}{x(\sqrt{x-1}-1)} = -\frac{1}{4}ln|x-1+1| + \frac{1}{2}tan^{-1}\left(\sqrt{x-1}\right) + \frac{1}{2}ln|\sqrt{x-1}-1| + C$$

$$\int \frac{dx}{x(\sqrt{x-1}-1)} = -\frac{1}{4}ln|x| + \frac{1}{2}tan^{-1}\left(\sqrt{x-1}\right) + \frac{1}{2}ln|\sqrt{x-1}-1| + C$$

$$\int \frac{dx}{x(\sqrt{x-1}-1)} = \frac{1}{2}\left[-\frac{1}{2}ln|x| + tan^{-1}\left(\sqrt{x-1}\right) + ln|\sqrt{x-1}-1|\right] + C$$

$$\int \frac{dx}{x(\sqrt{x-1}-1)} = \frac{1}{2}\left[ln|\sqrt{x-1}-1| - \frac{1}{2}ln|x| + tan^{-1}\left(\sqrt{x-1}\right)\right] + C$$

$$2\int \frac{dx}{x(\sqrt{x-1}-1)} = 2*\frac{1}{2}\left[ln|\sqrt{x-1}-1| - \frac{1}{2}ln(x) + tan^{-1}\left(\sqrt{x-1}\right)\right] + C$$

$$\boxed{\int \frac{dx}{x(\sqrt{x-1}-1)} = ln\left|\frac{\sqrt{x-1}-1}{\sqrt{x}}\right| + tan^{-1}\left(\sqrt{x-1}\right) + C}$$

218

$$\int \frac{(x+3)dx}{(x+6)\sqrt{x+2}}$$

The change of variable is

$$x + 2 = z^2 \;\to\; x = z^2 - 2 \;\to\; dx = 2z\, dz\,, z = \sqrt{x+2}$$

$$\int \frac{(x+3)dx}{(x+6)\sqrt{x+2}} = \int \frac{(z^2 - 2 + 3)2z\, dz}{(z^2 - 2 + 6)\, z}$$

$$\int \frac{(x+3)dx}{(x+6)\sqrt{x+2}} = 2 \int \frac{(z^2 + 1)\, dz}{(z^2 + 4)}$$

$$\int \frac{(x+3)dx}{(x+6)\sqrt{x+2}} = 2 \int \left(1 - \frac{3}{z^2 + 4}\right) dz$$

$$\int \frac{(x+3)dx}{(x+6)\sqrt{x+2}} = 2 \int dz - 6 \int \frac{dz}{z^2 + 2^2} = 2z - 3tan^{-1}\left(\frac{z}{2}\right) + C$$

$$\int \frac{(x+3)dx}{(x+6)\sqrt{x+2}} = 2\sqrt{x+2} - 3tan^{-1}\left(\frac{\sqrt{x+2}}{2}\right) + C$$

$$\boxed{\int \frac{(x+3)dx}{(x+6)\sqrt{x+2}} = 2\sqrt{x+2} - 3tan^{-1}\left(\frac{\sqrt{x+2}}{2}\right) + C}$$

219

$$\int \sqrt{1 + \sqrt{x}}\; dx$$

The irrational expression taken into account is: $\sqrt{x}$. Therefore, you have

$$x = z^2 \;\to\; dx = 2zdz\;,\; \sqrt{x} = z$$

$$\int \sqrt{1 + \sqrt{x}}\; dx = \int \sqrt{1 + z}\; 2z\, dz = 2 \int \sqrt{1 + z}\; z\, dz$$

Now we proceed to perform the following change of variable

$$1 + z = t^2 \quad \rightarrow \quad z = t^2 - 1 \rightarrow \quad dz = 2t\, dt \ , \ t = \sqrt{1+z}$$

$$2\int \sqrt{1+z}\ z\, dz = 2\int t\,(t^2-1)2t\, dt = 4\int (t^4 - t^2)dt$$

$$2\int \sqrt{1+z}\ z\, dz = 4\int t^4 dt - 4\int t^2 dt$$

$$2\int \sqrt{1+z}\ z\, dz = \frac{4}{5}t^5 - \frac{4}{3}t^3 + C = \frac{4}{5}\left(\sqrt{1+z}\right)^5 - \frac{4}{3}\left(\sqrt{1+z}\right)^3 + C$$

As $\sqrt{x} = z$

$$\int \sqrt{1+\sqrt{x}}\ dx = \frac{4}{5}(1+\sqrt{x})^{\frac{5}{2}} - \frac{4}{3}(1+\sqrt{x})^{\frac{3}{2}} + C$$

$$\boxed{\int \sqrt{1+\sqrt{x}}\ dx = \frac{4}{5}(1+\sqrt{x})^{\frac{5}{2}} - \frac{4}{3}(1+\sqrt{x})^{\frac{3}{2}} + C}$$

$$\int \sqrt{\frac{1+x}{1-x}}\ dx \qquad\qquad 220$$

The change of variable is

$$\frac{1+x}{1-x} = z^2 \rightarrow 1+x = z^2(1-x) \rightarrow 1+x = z^2 - xz^2 \rightarrow x + xz^2 = z^2 - 1$$

$$\rightarrow x(1+z^2) = z^2 - 1 \rightarrow x = \frac{z^2-1}{z^2+1}$$

Then you derive both members of the equation

$$dx = \frac{(z^2+1)2z - (z^2-1)2z}{(z^2+1)^2}dz \rightarrow dx = \frac{4z\, dz}{(z^2+1)^2}\ , \quad z = \sqrt{\frac{1+x}{1-x}}$$

$$\int \sqrt{\frac{1+x}{1-x}} \; dx = \int z \; \frac{4z \, dz}{(z^2+1)^2} = 4\int \frac{z^2 \, dz}{(z^2+1)^2}$$

It is expressed $(z^2+1)^2$ in terms of a root to then apply the trigonometric substitution integration technique.

$$\int \sqrt{\frac{1+x}{1-x}} \; dx = 4\int \frac{z^2 \, dz}{\left(\sqrt{z^2+1}\right)^4}$$

The form is: $a^2 + u^2$

Where $z = \tan\theta \;\; \rightarrow \;\; dz = \sec^2\theta \, d\theta \, , \; \sqrt{z^2+1} = \sec\theta$

$$4\int \frac{z^2 \, dz}{\left(\sqrt{z^2+1}\right)^4} = 4\int \frac{\tan^2\theta \sec^2\theta \, d\theta}{\sec^4\theta} = 4\int \frac{\tan^2\theta}{\sec^2\theta} \, d\theta = 4\int \frac{\frac{sen^2\theta}{\cos^2\theta}}{\frac{1}{\cos^2\theta}} \, d\theta$$

$$4\int \frac{z^2 \, dz}{\left(\sqrt{z^2+1}\right)^4} = 4\int sen^2\theta \, d\theta = 4\int \frac{1-\cos 2\theta}{2} \, d\theta$$

$$4\int \frac{z^2 \, dz}{\left(\sqrt{z^2+1}\right)^4} = 2\int d\theta - 2\int \cos 2\theta \, d\theta = 2\theta - 2sen 2\theta + C$$

Now the triangle is constructed to return to the original variable

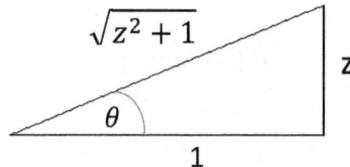

$$\tan\theta = z \;\; \rightarrow \;\; \theta = \tan^{-1}(z), \; sen 2\theta = 2 sen\theta \cos\theta, \; sen\theta = \frac{z}{\sqrt{z^2+1}}, \cos\theta = \frac{1}{\sqrt{z^2+1}}$$

$$4\int \frac{z^2 \, dz}{\left(\sqrt{z^2+1}\right)^4} = 2\theta - sen 2\theta + C = 2\theta - 2 sen\theta \cos\theta + C$$

$$4 \int \frac{z^2 \, dz}{\left(\sqrt{z^2+1}\right)^4} = 2\tan^{-1}(z) - 2\frac{z}{\sqrt{z^2+1}}\frac{1}{\sqrt{z^2+1}} = 2\tan^{-1}(z) - \frac{2z}{z^2+1} + C$$

$$\int \sqrt{\frac{1+x}{1-x}} \; dx = 2\tan^{-1}\left(\sqrt{\frac{1+x}{1-x}}\right) - \frac{2\sqrt{\frac{1+x}{1-x}}}{\frac{1+x}{1-x}+1} + C$$

$$\boxed{\int \sqrt{\frac{1+x}{1-x}} \; dx = 2\tan^{-1}\left(\sqrt{\frac{1+x}{1-x}}\right) - (1-x)\sqrt{\frac{1+x}{1-x}} + C}$$

$$\int \frac{\sqrt[3]{x} \, dx}{1 + \sqrt[3]{x^2}}$$

221

There are two irrational expressions of the same index. It proceeds to make the change of variable

$$x = z^3 \quad \rightarrow \quad dx = 3z^2 dz \;, \quad \sqrt[3]{x} = \sqrt[3]{z^3} \quad \rightarrow \quad \sqrt[3]{x} = z$$

Consequently, you have

$$\int \frac{\sqrt[3]{x}\,dx}{1+\sqrt[3]{x^2}} = \int \frac{z\,3z^2 dz}{1+\sqrt[3]{(z^3)^2}} = \int \frac{3z^3 dz}{1+\sqrt[3]{z^6}} = 3\int \frac{z^3 dz}{1+z^2}$$

As the degree of the polynomial of the numerator is greater than the degree of the denominator, the division of the polynomials is made

$$\frac{z^3}{z^2+1} = z - \frac{z}{z^2+1}$$

$$3\int \frac{z^3 dz}{z^2+1} = 3\int z\,dz - 3\int \frac{z\,dz}{z^2+1} \rightarrow u = z^2+1 \rightarrow du = 2z\,dz$$

$$3 \int \frac{z^3 dz}{z^2 + 1} = \frac{3}{2} z^2 - \frac{3}{2} \ln|u| + C = \frac{3}{2} z^2 - \frac{3}{2} \ln|z^2 + 1| + C$$

$$\boxed{\int \frac{\sqrt[3]{x}\, dx}{1 + \sqrt[3]{x^2}} = \frac{3}{2} \sqrt[3]{x^2} - \frac{3}{2} \ln \left| \sqrt[3]{x^2} + 1 \right| + C}$$

222 $\qquad\qquad\qquad\qquad\qquad\qquad\qquad$ $\boxed{\int \frac{\sqrt{x}\, dx}{\sqrt[4]{x^3} + 1}}$

The least common multiple of the indexes of roots 2 and 4 is determined. The m.c.m. is 4
The variable change is made

$$x = z^4 \rightarrow dx = 4z^3 dz \ , \ \sqrt[4]{x} = z$$

$$\int \frac{\sqrt{x}\, dx}{\sqrt[4]{x^3} + 1} = \int \frac{\sqrt[4]{x^2}\, dx}{\sqrt[4]{x^3} + 1} = \int \frac{\sqrt[4]{z^8}\, 4z^3 dz}{\sqrt[4]{z^{12}} + 1} = \int \frac{z^2\, 4z^3 dz}{z^3 + 1}$$

$$\int \frac{\sqrt{x}\, dx}{\sqrt[4]{x^3} + 1} = 4 \int \frac{z^5\, dz}{z^3 + 1}$$

Next, we proceed to perform the division of the polynomials

$$\frac{z^5}{z^3 + 1} = z^2 - \frac{z^2}{z^3 + 1}$$

$$\int \frac{\sqrt{x}\, dx}{\sqrt[4]{x^3} + 1} = 4 \int z^2\, dz - 4 \int \frac{z^2 dz}{z^3 + 1} \ \rightarrow \ u = z^3 + 1, du = 3z^2 dz$$

$$\int \frac{\sqrt{x}\, dx}{\sqrt[4]{x^3} + 1} = \frac{4}{3} z^3 - \frac{4}{3} \int \frac{du}{u} = \frac{4}{3} z^3 - \frac{4}{3} \ln|z^3 + 1| + C$$

$$\boxed{\int \frac{\sqrt{x}\, dx}{\sqrt[4]{x^3} + 1} = \frac{4}{3} \sqrt[4]{x^3} - \frac{4}{3} \ln \left(\sqrt[4]{x^3} + 1 \right) + C}$$

$$\int \frac{dx}{\sqrt{\sqrt{x}+1}}$$

223

The next change of variable is made.

$$\sqrt{x}+1 = z^2 \rightarrow \sqrt{x} = z^2 - 1 \rightarrow x = (z^2-1)^2,\; dx = 2(z^2-1)2z\,dz \rightarrow$$
$$dx = (4z^3 - 4z)dz \rightarrow \text{To find the dx the rule of the chain was applied.}$$

You can also change the variable $x = z^2$

$$\int \frac{dx}{\sqrt{\sqrt{x}+1}} = \int \frac{(4z^3-4z)dz}{\sqrt{z^2}} = \int \frac{(4z^3-4z)dz}{z}$$

$$\int \frac{dx}{\sqrt{\sqrt{x}+1}} = 4\int \frac{z^3}{z}\,dz - 4\int \frac{z}{z}\,dz$$

$$\int \frac{dx}{\sqrt{\sqrt{x}+1}} = 4\int z^2\,dz - 4\int dz$$

$$\int \frac{dx}{\sqrt{\sqrt{x}+1}} = \frac{4}{3}z^3 - 4z + C$$

$$\int \frac{dx}{\sqrt{\sqrt{x}+1}} = \frac{4}{3}\left(\sqrt{\sqrt{x}+1}\right)^3 - 4\sqrt{\sqrt{x}+1} + C$$

$$\boxed{\int \frac{dx}{\sqrt{\sqrt{x}+1}} = \frac{4}{3}\sqrt{(\sqrt{x}+1)^3} - 4\sqrt{\sqrt{x}+1} + C}$$

224

$$\int \frac{(\sqrt{x+1}+2)dx}{(x+1)^2 - \sqrt{x+1}}$$

The next change of variable is made.

$$x + 1 = t^2 \rightarrow x = t^2 - 1 \rightarrow dx = 2t\,dt\ ,\sqrt{x+1} = t$$

Therefore, you have

$$\int \frac{(\sqrt{x+1}+2)dx}{(x+1)^2 - \sqrt{x+1}} = \int \frac{(t+2)\,2t\,dt}{(t^2-1+1)^2 - t} = 2\int \frac{(t^2+2t)\,dt}{t^4 - t}$$

$$\int \frac{(\sqrt{x+1}+2)dx}{(x+1)^2 - \sqrt{x+1}} = 2\int \frac{t(t+2)\,dt}{t(t^3-1)} = 2\int \frac{(t+2)\,dt}{t^3 - 1}$$

$$\int \frac{(\sqrt{x+1}+2)dx}{(x+1)^2 - \sqrt{x+1}} = 2\int \frac{(t+2)\,dt}{(t-1)(t^2+t+1)}$$

The resulting integral is solved by applying the technique of partial fraction integration.

$$\frac{(t+2)}{(t-1)(t^2+t+1)} = \frac{A}{t-1} + \frac{Bt+C}{t^2+t+1}$$

$$t + 2 = A(t^2+t+1) + (Bt+C)(t-1)$$

$$t + 2 = At^2 + At + A + Bt^2 - Bt + Ct - C$$

$$t + 2 = (A+B)t^2 + (A-B+C)t + (A-C)$$

With the above information the system of equations is constructed.

$$
\begin{array}{llll}
A & +B & & = 0 & \text{Ec. } 1 \\
A & -B & +C & = 1 & \text{Ec. } 2 \\
A & & -C & = 2 & \text{Ec. } 3
\end{array}
$$

Eq. 1 and Eq. 2 are combined and obtained

$$A \;+\; B \qquad = 0$$
$$A \;-\; B \;+\; C \;=\; 1 \qquad \rightarrow \qquad 2A + C = 1$$

This result is combined with Eq. 3

$$A \;-\; C \qquad = 2$$
$$2A + C \qquad = 1 \qquad \rightarrow \qquad 3A = 3 \qquad \rightarrow \qquad \boldsymbol{A = 1}$$

The previous result is combined with Ec. 1

$$A \;+\; B \qquad = 0 \quad \rightarrow \quad 1 + B = 0 \quad \rightarrow \quad \boldsymbol{B = -1}$$

The previous result is combined with Ec.3

$$A \;-\; C = 2 \quad \rightarrow \quad 1 - C = 2 \quad \rightarrow \quad \boldsymbol{C = -1}$$

Consequently, you have

$$2\int \frac{(t+2)\;dt}{(t-1)(t^2+t+1)} = 2\int \frac{A}{t-1}dt + 2\int \frac{Bt+C}{t^2+t+1}dt$$

$$2\int \frac{(t+2)\;dt}{(t-1)(t^2+t+1)} = 2\int \frac{dt}{t-1} - 2\int \frac{(t+1)dt}{t^2+t+1}$$

$$2\int \frac{(t+2)\;dt}{(t-1)(t^2+t+1)} = 2\int \frac{dt}{t-1} - 2\int \frac{t\;dt}{t^2+t+1} - 2\int \frac{dt}{t^2+t+1}$$

$$2\int \frac{(t+2)\;dt}{(t-1)(t^2+t+1)} = 2\int \frac{dt}{t-1} - \int \frac{(2t+1-1)\;dt}{t^2+t+1} - 2\int \frac{dt}{t^2+t+1}$$

$$2\int \frac{(t+2)\;dt}{(t-1)(t^2+t+1)} = 2\int \frac{dt}{t-1} - \int \frac{(2t+1)\;dt}{t^2+t+1} + \int \frac{dt}{t^2+t+1} - 2\int \frac{dt}{t^2+t+1}$$

$$2\int \frac{(t+2)\;dt}{(t-1)(t^2+t+1)} = 2\int \frac{dt}{t-1} - \int \frac{(2t+1)\;dt}{t^2+t+1} - \int \frac{dt}{t^2+t+1}$$

$$2\int \frac{(t+2)\;dt}{(t-1)(t^2+t+1)} = 2\int \frac{dt}{t-1} - \int \frac{(2t+1)\;dt}{t^2+t+1} - \int \frac{dt}{\left(t+\frac{1}{2}\right)^2 + \left(\frac{\sqrt{3}}{2}\right)^2}$$

Note: In the penultimate step, in the last integral, completion of squares was applied

287

$$2\int \frac{(t+2)\ dt}{(t-1)(t^2+t+1)} = 2ln|t-1| - ln|t^2+t+1| - \frac{1}{\frac{\sqrt{3}}{2}}tan^{-1}\left(\frac{t+\frac{1}{2}}{\frac{\sqrt{3}}{2}}\right) + C$$

$$2\int \frac{(t+2)\ dt}{(t-1)(t^2+t+1)} = 2ln|t-1| - ln|t^2+t+1| - \frac{2}{\sqrt{3}}tan^{-1}\left(\frac{\frac{2t+1}{2}}{\frac{\sqrt{3}}{2}}\right) + C$$

$$2\int \frac{(t+2)\ dt}{(t-1)(t^2+t+1)} = 2ln|t-1| - ln|t^2+t+1| - \frac{2\sqrt{3}}{3}tan^{-1}\left(\frac{4t+2}{2\sqrt{3}}\right) + C$$

$$2\int \frac{(t+2)\ dt}{(t-1)(t^2+t+1)} = 2ln|t-1| - ln|t^2+t+1| - \frac{2\sqrt{3}}{3}tan^{-1}\left(\frac{2t+1}{\sqrt{3}}\right) + C$$

As $\sqrt{x+1} = t$ is obtained

$$\int \frac{(\sqrt{x+1}+2)dx}{(x+1)^2-\sqrt{x+1}} = ln\left|\frac{(\sqrt{x+1}-1)^2}{(\sqrt{x+1})^2+\sqrt{x+1}+1}\right| - \frac{2\sqrt{3}}{3}tan^{-1}\left(\frac{2\sqrt{x+1}+1}{\sqrt{3}}\right) + C$$

$$\boxed{\int \frac{(\sqrt{x+1}+2)dx}{(x+1)^2-\sqrt{x+1}} = ln\left|\frac{(\sqrt{x+1}-1)^2}{\sqrt{x+1}+x+2}\right| - \frac{2\sqrt{3}}{3}\ tan^{-1}\left[\frac{1}{3}(2\sqrt{x+1}+1)\sqrt{3}\right] + C}$$

Strategies to start the resolution of indefinite integrals

$$\int (2sen\ x + 3\cos x)\ dx$$

225

Note that the integrand is composed of the sum of two trigonometric functions.
In the theoretical foundation Theorem 1 says

$$If\ \ F'(x) = f(x)\ and\ G'(x) = g(x), then,$$

$$\int [f(x) \pm g(x)]dx = \int f(x)dx \pm \int g(x)dx = F(x) \pm G(x) + C$$

Therefore, you have:

$$\int (2sen\ x + 3\cos x)\ dx = \int 2sen\ x\ dx + \int 3\cos x\ dx$$

For both integrals the constant multiple rule applies

$$\int (2sen\ x + 3\cos x)\ dx = 2\int sen\ x\ dx + 3\int \cos x\ dx$$

Both the first integral and the second integral are immediate. (See table of integrals)

$$\int (2sen\ x + 3\cos x)\ dx = -2\cos x + 3sen\ x + C$$

$$\boxed{\int (2sen\ x + 3\cos x)\ dx = -2\cos x + 3sen\ x + C}$$

$$\int \frac{e^t - 1}{e^t + 1}\ dt$$

226

Let's start solving the integral by making the following variable change: $u = e^t + 1$ And then there is your differential of u. $du = e^t\ dt$. Note that in the differential of u there is   no direct correspondence with the original integral, that is, an $e^t\ dt$. This situation suggests us to rewrite the integral in the following terms

$$\int \frac{e^t - 1}{e^t + 1} \, dt = \int \frac{e^t}{e^t + 1} \, dt - \int \frac{1}{e^t + 1} \, dt$$

Note that the first integral can be resolved by applying the previous variable change; $u = e^t + 1$. So you have

$$\int \frac{e^t - 1}{e^t + 1} \, dt = \int \frac{du}{u} - \int \frac{1}{e^t + 1} \, dt$$

The other integral is solved by applying the following mathematical artifice: The integrand is multiplied and divided by e^{-t}.

$$\int \frac{dt}{e^t + 1} = \int \frac{e^{-t} \, dt}{e^{-t}(e^t + 1)} = \int \frac{e^{-t} \, dt}{e^{-t} + 1}$$

The resulting integral is solved by applying the following variable change: $w = e^{-t} + 1$ $dw = -e^{-t} dt$.

As a result, you get

$$\int \frac{e^t - 1}{e^t + 1} \, dt = \int \frac{du}{u} - \int \frac{dt}{e^t + 1} = \int \frac{du}{u} + \int \frac{dw}{w} = ln|u| + ln|w| + C$$

$$\boxed{\int \frac{e^t - 1}{e^t + 1} \, dt = ln[(e^t + 1)(e^{-t} + 1)] + C}$$

227

$$\int \frac{e^x \, dx}{\sqrt{1 - e^{2x}}}$$

The importance of rewriting the integral problem is that it allows us to simplify the integral. For the simplification it is necessary to apply algebraic procedures known as mathematical devices such as: Radiation properties, potentiation properties, common factor, multiply and divide the integrand by a factor.

In the integral proposal, potentiation properties are applied in the radical

$$\sqrt{1 - e^{2x}} = \sqrt{1 - (e^x)^2}$$

The integral is rewritten

$$\int \frac{e^x \, dx}{\sqrt{1 - e^{2x}}} = \int \frac{e^x \, dx}{\sqrt{1 - (e^x)^2}}$$

Note that in the simplified integral the following variable change can be made

$$u = e^x \longrightarrow du = e^x \, dx$$

Since there is a direct correspondence between the du and the numerator of the integrand, we have

$$\int \frac{e^x \, dx}{\sqrt{1 - (e^x)^2}} = \int \frac{du}{\sqrt{1 - u^2}}$$

The resulting integral is immediate and is of the form

$$\int \frac{du}{\sqrt{1 - u^2}} = arcsen \, u + C$$

As a result, you get

$$\boxed{\int \frac{e^x \, dx}{\sqrt{1 - e^{2x}}} = arcsen \, (e^x) + C}$$

$$\boxed{\int \frac{dt}{3^t + 4}}$$

228

When changing the variable $u = 3^t + 4$ and then deriving it is: $du = 3^t \, ln3dt$, it can be seen that in the du the term 3^t, appears, which does not appear in the numerator of the integrand. This situation tells us that the denominator of the integrand must be built in the numerator of the same. To achieve the above, the following mathematical artifice applies. Multiply and divide by 4 the integral

$$\int \frac{dt}{3^t + 4} = \frac{1}{4} \int \frac{4 \, dt}{3^t + 4}$$

Then the term is added and subtracted 3^t

$$\int \frac{dt}{3^t + 4} = \frac{1}{4}\int \frac{4\,dt}{3^t + 4} = \frac{1}{4}\int \frac{4 + 3^t - 3^t}{3^t + 4}\,dt$$

Next, the resulting integral is rewritten by separating it into two integrals

$$\frac{1}{4}\int \frac{4 + 3^t - 3^t}{3^t + 4}\,dt = \frac{1}{4}\int \frac{3^t + 4}{3^t + 4}\,dt - \frac{1}{4}\int \frac{3^t\,dt}{3^t + 4}$$

The result of the first integral is immediate: $\frac{1}{4}t + c_1$ and that of the second integral is obtained by applying the variable change: $u = 3^t + 4 \;\rightarrow\; du = 3^t \ln3\,dt$. Observe the reader that in the differential of (u) the factor ln3 appears multiplying, therefore, the same factor divides the integral.

$$\boxed{\int \frac{dt}{3^t + 4} = \frac{1}{4}t - \frac{\ln|3^t + 4|}{4\ln3} + C}$$

229 $\qquad\qquad \int e^{sen^2 x}\, sen\,2x\,dx$

Notice that the square sine function its angle is x and the angle of the other trigonometric function is 2x. This situation suggests the application of a trigonometric identity that allows us to unify the angles. The identity is

$$sen^2 x = \frac{1 - \cos 2x}{2}$$

Then we proceed to rewrite the integral problem

$$\int e^{sen^2 x}\, sen\,2x\,dx = \int e^{\frac{1 - \cos 2x}{2}}\, sen\,2x\,dx$$

Now the next variable change is made

$$Be:\ u = \frac{1 - \cos 2x}{2} \;\rightarrow\; u = \frac{1}{2}(1 - \cos 2x) \;\rightarrow\; du = sen\,2x\,dx$$

Therefore, you have

$$\int e^{\frac{1-\cos 2x}{2}} sen\, 2x\, dx = \int e^u\, du$$

$$\int e^{\frac{1-\cos 2x}{2}} sen\, 2x\, dx = e^u + C = e^{\frac{1-\cos 2x}{2}} + C$$

$$\boxed{\int e^{sen^2 x}\, sen\, 2x\, dx = e^{sen^2 x} + C}$$

$$\int \sqrt{\frac{ln(x + \sqrt{x^2+1})}{1+x^2}}\, dx$$

230

It is important that the reader be aware of the significance of the rewriting of the problem integrals. Since this procedure simplifies the integral and determines which integration technique is going to be used.

In the integrand we find a root which contains a quotient in its subradical quantity. This information suggests the idea of applying radiation properties to simplify the integral. That is, separate the quotient into two radicals.

$$\int \sqrt{\frac{ln(x + \sqrt{x^2+1})}{1+x^2}}\, dx = \int \frac{\sqrt{ln(x + \sqrt{x^2+1})}}{\sqrt{1+x^2}}\, dx$$

Thanks to the simplification of the integral it can be seen that in the integrand we have an una expresión $\frac{dx}{\sqrt{1+x^2}}$ which gives us the idea of making the next variable change.

$$u = ln\left(x + \sqrt{x^2+1}\right) \quad \rightarrow \quad du = \frac{dx}{\sqrt{x^2+1}}$$

Note: The algebraic procedure to find (du) is found in exercise number 27 of this book.

As a result, you get

$$\int \frac{\sqrt{ln(x + \sqrt{x^2+1})}}{\sqrt{1+x^2}}\, dx = \int \sqrt{u}\, du = \int u^{\frac{1}{2}}\, du = \frac{u^{\frac{3}{2}}}{\frac{3}{2}} + c$$

$$= \frac{2}{3}\sqrt{u^3} + c = \frac{2}{3}\sqrt{\left[\ln(x + \sqrt{x^2 + 1})\right]^3} + C$$

$$\boxed{\int \sqrt{\frac{\ln(x + \sqrt{x^2 + 1})}{1 + x^2}}\, dx = \frac{2}{3}\sqrt{\left[\ln(x + \sqrt{x^2 + 1})\right]^3} + C}$$

231

$$\boxed{\int \frac{x^3 - x + 3}{x^2 + x - 2}\, dx}$$

The reader should bear in mind the importance of describing the integral proposal. What does the above mean? Well, in our case with the integral problem, it is observed that the integrand is the quotient of two polynomials. That the degree of the polynomial of the numerator is greater than the degree of the polynomial of the denominator. That is, we are in the presence of a rational integral. Now we proceed to solve the integral.

How the degree of the polynomial of the numerator is greater than the degree of the polynomial of the denominator is the division of the polynomials with the purpose of simplifying the integral.

$$\frac{x^3 - x + 3}{x^2 + x - 2} = x - 1 + \frac{2x + 1}{x^2 + x - 2}$$

Consequently, you have

$$\int \frac{x^3 - x + 3}{x^2 + x - 2}\, dx = \int x\, dx - \int dx + \int \frac{2x + 1}{x^2 + x - 2}\, dx$$

In the previous partial result the first and second integral are immediate. The third integral is resolved with a change of variable. $u = x^2 + x - 2 \;\rightarrow\; du = (2x + 1)dx$
Consequently, you have

$$\int \frac{x^3 - x + 3}{x^2 + x - 2}\, dx = \frac{1}{2}x^2 - x + \int \frac{du}{u}$$

$$\boxed{\int \frac{x^3 - x + 3}{x^2 + x - 2}\, dx = \frac{1}{2}x^2 - x + \ln|x^2 + x - 2| + C}$$

$$\int \frac{x^2}{x^4 - 2x^2 - 8} dx$$

232

Note that the degree of the polynomial of the numerator is smaller than the degree of the polynomial of the denominator, therefore, the division of the polynomials can not be performed in order to simplify the integral. In this case, we proceed to factor the polynomial of the denominator.

To factor the polynomial of the denominator, it is first rewritten:

$$x^4 - 2x^2 - 8 = (x^2)^2 - 2(x^2) - 8$$

The reader may appreciate that the result refers to a case of factoring the form $x^2 + bx + c$. To develop this case we proceed to find two numbers that multiplied den - 8 and that subtracted den - 2.

$$(x^2)^2 - 2(x^2) - 8 = (x^2 - 4)(x^2 + 2)$$

When solving the square difference of the first parenthesis the factorization is

$$x^4 - 2x^2 - 8 = (x - 2)(x + 2)(x^2 + 2)$$

Consequently, you have

$$\int \frac{x^2}{x^4 - 2x^2 - 8} dx = \int \frac{x^2}{(x - 2)(x + 2)(x^2 + 2)} dx$$

The resulting integral is solved by applying the technique of partial fraction integration. Then

$$\frac{x^2}{(x - 2)(x + 2)(x^2 + 2)} = \frac{A}{(x - 2)} + \frac{B}{(x + 2)} + \frac{Cx + D}{(x^2 + 2)}$$

$$x^2 = A(x + 2)(x^2 + 2) + B(x - 2)(x^2 + 2) + (Cx + D)(x + 2)(x - 2)$$

$$x^2 = Ax^3 + 2Ax + 2Ax^2 + 4A + Bx^3 + 2Bx - 2Bx^2 - 4B + Cx^3 - 4Cx + Dx^2 - 4D$$

$$x^2 = (A + B + C)x^3 + (2A - 2B + D)x^2 + (2A + 2B - 4C)x + (4A - 4B - 4D)$$

$$\begin{array}{llll} A & + B & + C & = 0 & \text{Equation 1} \\ -2A & + -2B & & + D = 1 & \text{Equation 2} \\ 2A & + 2B & - 4C & = 0 & \text{Equation 3} \\ 4A & - 4B & & - 4D = 0 & \text{Equation 4} \end{array}$$

Solving the system of equations we have the values for A, B, C and D.

$$A = \frac{1}{6}, \quad B = -\frac{1}{6}, \quad C = 0 \quad D = \frac{1}{3}$$

$$\int \frac{x^2}{x^4 - 2x^2 - 8} dx = \int \left[\frac{A}{(x-2)} + \frac{B}{(x+2)} + \frac{Cx+D}{(x^2+2)} \right] dx$$

$$\int \frac{x^2}{x^4 - 2x^2 - 8} dx = \int \frac{A}{(x-2)} dx + \int \frac{B}{(x+2)} dx + \int \frac{Cx+D}{(x^2+2)} dx$$

$$\int \frac{x^2}{x^4 - 2x^2 - 8} dx = \int \frac{\frac{1}{6}}{(x-2)} dx + \int \frac{-\frac{1}{6}}{(x+2)} dx + \int \frac{\frac{1}{3}}{(x^2+2)} dx$$

$$\int \frac{x^2}{x^4 - 2x^2 - 8} dx = \frac{1}{6} \int \frac{dx}{(x-2)} - \frac{1}{6} \int \frac{dx}{(x+2)} + \frac{1}{3} \int \frac{dx}{(x^2+2)}$$

$$\int \frac{x^2}{x^4 - 2x^2 - 8} dx = \frac{1}{6} \left[\int \frac{dx}{(x-2)} - \int \frac{dx}{(x+2)} + 2 \int \frac{dx}{(x^2+2)} \right]$$

$$\int \frac{x^2}{x^4 - 2x^2 - 8} dx = \frac{1}{6} \left[ln|(x-2)| - ln|(x+2)| + \sqrt{2} arctan \frac{x}{\sqrt{2}} \right] + c$$

$$\int \frac{x^2}{x^4 - 2x^2 - 8} dx = \frac{1}{6} \left[ln \frac{(x-2)}{(x+2)} + \sqrt{2} arctan \left(\frac{x}{\sqrt{2}} \right) \right] + c$$

$$\boxed{\int \frac{x^2}{x^4 - 2x^2 - 8} dx = \frac{1}{6} \left[ln \frac{(x-2)}{(x+2)} + \sqrt{2} arctan \left(\frac{1}{2} x\sqrt{2} \right) \right] + C}$$

$$\int \frac{sen\,\theta\,d\theta}{cos^2 + cos\theta - 2}$$

Note that the denominator of the fraction is a second-degree trigonometric equation that can be factored.

The denominator of the fraction is factored:

$$cos^2 + cos\theta - 2 = (cos\theta + 2)(cos\theta - 1)$$

$$\int \frac{sen\,\theta\,d\theta}{cos^2 + cos\theta - 2} = \int \frac{sen\,\theta\,d\theta}{(cos\theta + 2)(cos\theta - 1)}$$

In the resulting integral, the following variable change is made.

$$u = cos\theta + 2 \quad \rightarrow \quad du = -sen\theta\,d\theta$$

$$if\ cos\theta = u - 2 \quad then\ cos\theta - 1 = u - 2 - 1 \quad \rightarrow \quad cos\theta - 1 = u - 3$$

$$\int \frac{sen\,\theta\,d\theta}{cos^2 + cos\theta - 2} = -\int \frac{du}{u(u - 3)}$$

Now the integrand is broken down into partial fractions and mathematically operated:

$$\frac{1}{u(u - 3)} = \frac{A}{u} + \frac{B}{(u - 3)}$$

$$1 = A(u - 3) + B(u)$$

$$1 = Au - 3A + Bu$$

$$1 = (A + B)u - 3A$$

$$-3A = 1 \quad \rightarrow \quad A = -\frac{1}{3}$$

$$A + B = 0 \quad \rightarrow \quad B = \frac{1}{3}$$

Consequently, you have

$$-\int \frac{du}{u(u-3)} = -\left(-\frac{1}{3}\int \frac{du}{u} + \frac{1}{3}\int \frac{du}{u-3}\right)$$

$$\int \frac{sen\,\theta\,d\theta}{cos^2 + cos\theta - 2} = \frac{1}{3}ln|cos\theta + 2| - \frac{1}{3}ln|cos\theta + 2 - 3| + C$$

$$\int \frac{sen\,\theta\,d\theta}{cos^2 + cos\theta - 2} = \frac{1}{3}ln|cos\theta + 2| - \frac{1}{3}ln|cos\theta - 1| + C$$

Because the $|cos\theta| < 1$ has:

$$|cos\theta - 1| = |1 - cos\theta|$$

Therefore, you get:

$$\int \frac{sen\,\theta\,d\theta}{cos^2 + cos\theta - 2} = \frac{1}{3}ln|2 + cos\theta| - \frac{1}{3}ln|1 - cos\theta| + C$$

$$\int \frac{sen\,\theta\,d\theta}{cos^2 + cos\theta - 2} = \frac{1}{3}ln\left|\frac{2 + cos\theta}{1 - cos\theta}\right| + C$$

$$\boxed{\int \frac{\boldsymbol{sen\,\theta\,d\theta}}{\boldsymbol{cos^2 + cos\theta - 2}} = \frac{1}{3}ln\left|\frac{2 + cos\theta}{1 - cos\theta}\right| + C}$$

234

$$\boxed{\int \frac{dx}{e^{2x} + e^x - 2}}$$

The authors of this book follow the students, as input to initiate the resolution of an integral is to apply the substitution or change of variable u and du. In the present integral the application of this mathematical resource is not very useful. Another alternative is to rewrite the integral in such a way that a second degree function appears. To then complete squares, factor and apply the technique of partial fraction integration.

$$e^{2x} + e^x - 2 = (e^x)^2 + e^x - 2 = (e^x)^2 + e^x + \frac{1}{4} - 2 - \frac{1}{4}$$

$$e^{2x} + e^x - 2 = \left[(e^x)^2 + e^x + \frac{1}{4}\right] - \frac{9}{4} = \left(e^x + \frac{1}{2}\right)^2 - \left(\frac{3}{2}\right)^2$$

$$\int \frac{dx}{e^{2x} + e^x - 2} = \int \frac{dx}{\left(e^x + \frac{1}{2}\right)^2 - \left(\frac{3}{2}\right)^2}$$

Now we proceed to transform the exponential term as a function of u and du.

Be $\qquad u = e^x + \frac{1}{2} \;\rightarrow\; du = e^x dx$

It clears $\quad dx \rightarrow dx = \frac{du}{e^x}$

As $\qquad e^x = u - \frac{1}{2}$

You have $\quad dx = \dfrac{du}{u - \frac{1}{2}}$

Then the substitution is made

$$\int \frac{dx}{\left(e^x + \frac{1}{2}\right)^2 - \left(\frac{3}{2}\right)^2} = \int \frac{\frac{du}{u - \frac{1}{2}}}{(u)^2 - \left(\frac{3}{2}\right)^2} = \int \frac{du}{\left(u - \frac{1}{2}\right)\left(u + \frac{3}{2}\right)\left(u - \frac{3}{2}\right)}$$

Now the integrand is broken down into simple fractions and mathematically operated

$$\frac{1}{\left(u - \frac{1}{2}\right)\left(u + \frac{3}{2}\right)\left(u - \frac{3}{2}\right)} = \frac{A}{\left(u - \frac{1}{2}\right)} + \frac{B}{\left(u + \frac{3}{2}\right)} + \frac{C}{\left(u - \frac{3}{2}\right)}$$

$$1 = A\left(u + \frac{3}{2}\right)\left(u - \frac{3}{2}\right) + B\left(u - \frac{1}{2}\right)\left(u - \frac{3}{2}\right) + C\left(u - \frac{1}{2}\right)\left(u + \frac{3}{2}\right)$$

Comment: A procedure to calculate the values of A, B and C is to substitute values suitable for (u) in the fundamental equation to obtain zeros in the other factors. That is to sayFor $u = \frac{1}{2}$ It is determined that the factors that cancel are B and C. Therefore,

$1 = A(2)(-1)$. Which implies that $A = -\frac{1}{2}$

For $u = -\dfrac{3}{2}$ It is determined that the factors that cancel are the A and C. Therefore,

$1 = B(-2)(-3)$. Which implies that $B = \dfrac{1}{6}$

For $u = \dfrac{3}{2}$ It is determined that the factors that are annulled are A and B. Therefore,

$1 = C(1)(3)$. Which implies that $C = \dfrac{1}{3}$

Consequently, you have

$$\int \frac{du}{\left(u-\frac{1}{2}\right)\left(u+\frac{3}{2}\right)\left(u-\frac{3}{2}\right)} = \int \frac{A\,du}{\left(u-\frac{1}{2}\right)} + \int \frac{B\,du}{\left(u+\frac{3}{2}\right)} + \int \frac{C\,du}{\left(u-\frac{3}{2}\right)}$$

$$\int \frac{du}{\left(u-\frac{1}{2}\right)\left(u+\frac{3}{2}\right)\left(u-\frac{3}{2}\right)} = -\frac{1}{2}\int \frac{du}{\left(u-\frac{1}{2}\right)} + \frac{1}{6}\int \frac{du}{\left(u+\frac{3}{2}\right)} + \frac{1}{3}\int \frac{du}{\left(u-\frac{3}{2}\right)}$$

$$= -\frac{1}{2}\ln\left|u-\frac{1}{2}\right| + \frac{1}{6}\ln\left|u+\frac{3}{2}\right| + \frac{1}{3}\ln\left|u-\frac{3}{2}\right| + C$$

$$= \frac{1}{6}\left[-3\ln\left|u-\frac{1}{2}\right| + \ln\left|u+\frac{3}{2}\right| + 2\ln\left|u-\frac{3}{2}\right|\right] + C$$

$$= \frac{1}{6}\ln\left|\frac{\left(u+\frac{3}{2}\right)\left(u-\frac{3}{2}\right)^2}{\left(u-\frac{1}{2}\right)^3}\right| + C$$

As $u = e^x + \dfrac{1}{2}$ and operating mathematically you have

$$\int \frac{dx}{e^{2x}+e^x-2} = \frac{1}{6}\ln\left|\frac{(e^x+2)(e^x-1)^2}{e^{3x}}\right| + C$$

$$\boxed{\int \frac{dx}{e^{2x}+e^x-2} = \frac{1}{6}\ln\left|\frac{(e^x+2)(e^x-1)^2}{e^{3x}}\right| + C}$$

$$\int \frac{\sec x}{\sec x + \tan x - 1}\,dx$$

235

The initial idea is to transform the trigonometric functions present in the integral in terms of sines and cosines. To then apply the technique of integration of functions rationales sines and cosines.

$$\int \frac{\sec x}{\sec x + \tan x - 1}\,dx = \int \frac{\dfrac{1}{\cos x}}{\dfrac{1}{\cos x} + \dfrac{sen\,x}{\cos x} - 1}\,dx$$

For the resolution of integrals containing sine and cosine functions the following substitutions will be made

$$sen\,x = \frac{2u}{1+u^2} \quad , \quad \cos x = \frac{1-u^2}{1+u^2} \quad , \quad dx = \frac{2}{1+u^2}\,du \quad , \quad u = tan\frac{x}{2}$$

$$\int \frac{\sec x}{\sec x + \tan x - 1}\,dx = \int \frac{\dfrac{1}{\cos x}}{\dfrac{1}{\cos x} + \dfrac{sen\,x}{\cos x} - 1}\,dx$$

$$\int \frac{\sec x}{\sec x + \tan x - 1}\,dx = \int \frac{\dfrac{1}{\cos x}}{\dfrac{1 + sen\,x - \cos x}{\cos x}}\,dx$$

$$\int \frac{\sec x}{\sec x + \tan x - 1}\,dx = \int \frac{\cos x}{\cos x(1 + sen\,x - \cos x)}\,dx$$

$$\int \frac{\sec x}{\sec x + \tan x - 1}\,dx = \int \frac{dx}{1 + sen\,x - \cos x}$$

$$\int \frac{\sec x}{\sec x + \tan x - 1}\,dx = \int \frac{\dfrac{2}{1+u^2}\,du}{1 + \dfrac{2u}{1+u^2} - \dfrac{1-u^2}{1+u^2}}$$

$$\int \frac{\sec x}{\sec x + \tan x - 1} \, dx = \int \frac{\dfrac{2}{1+u^2} \, du}{\dfrac{1+u^2+2u-1+u^2}{1+u^2}}$$

$$\int \frac{\sec x}{\sec x + \tan x - 1} \, dx = \int \frac{2}{2u(u+1)} \, du = \int \frac{1}{u(u+1)} \, du$$

Where the resulting integral is solved by applying partial fraction integration

$$\int \frac{1}{u(u+1)} \, du = \int \frac{A}{u} \, du + \int \frac{B}{u+1} \, du$$

$$\frac{1}{u(u+1)} = \frac{A}{u} + \frac{B}{u+1} \qquad \rightarrow \qquad 1 = A(u+1) + Bu$$

For $u = -1$ you have: $1 = -B \quad \rightarrow \quad B = -1$

For $u = 0$ you have: $1 = A \qquad \rightarrow \quad A = 1$

$$\int \frac{du}{u(u+1)} = \int \frac{du}{u} - \int \frac{du}{u+1}$$

$$\int \frac{du}{u(u+1)} = ln|u| - ln|u+1| + C$$

As $u = tan\dfrac{x}{2}$ results in

$$\int \frac{\sec x}{\sec x + \tan x - 1} \, dx = ln\left|tan\frac{x}{2}\right| - ln\left|tan\frac{x}{2}+1\right| + C$$

Applying logarithm properties

$$\boxed{\int \frac{\sec x}{\sec x + \tan x - 1} \, dx = ln\left|\frac{tan\dfrac{x}{2}}{tan\dfrac{x}{2}+1}\right| + C}$$

$$\int \frac{1 - senx}{(1 + senx)senx} dx$$

236

The integral problem can be solved by applying the technique of integration of rational functions of sines and cosines. If the respective substitutions are applied directly, laborious algebraic procedures can be created. The suggestion of the authors is to simplify the integral, that is, to rewrite it.

$$\int \frac{1 - senx}{(1 + senx)senx} dx = \int \left[\frac{(1 - sen\,x)}{(1 + sen\,x)} \frac{1}{sen\,x} \right] dx$$

Proceed to divide $(1 - sen\,x)$ with $(1 + sen\,x)$ to separate the integral problem into two integrals

$$\int \left[\frac{(1 - sen\,x)}{(1 + sen\,x)} \frac{1}{sen\,x} \right] dx = \int \left[\left(1 - \frac{2\,sen\,x}{1 + sen\,x} \right) \frac{1}{sen\,x} \right] dx$$

$$= \int \frac{dx}{sen\,x} - 2 \int \left(\frac{sen\,x}{1 + sen\,x} \right) \frac{1}{sen\,x} dx$$

$$= \int \frac{dx}{sen\,x} - 2 \int \frac{dx}{1 + sen\,x} \qquad \textbf{\textit{Equation 1}}$$

The first and second integral are solved

$$\int \frac{dx}{sen\,x} = \int \frac{\frac{2du}{1 + u^2}}{\frac{2u}{1 + u^2}} = \int \frac{du}{u} = ln|u| = ln\left| tan\frac{x}{2} \right| + C_1$$

$$2 \int \frac{dx}{1 + sen\,x} = 2 \int \frac{\frac{2du}{1 + u^2}}{1 + \frac{2u}{1 + u^2}} = 4 \int (u + 1)^{-2} du = -\frac{4}{tan\left(\frac{x}{2}\right) + 1} + C_2$$

Substituting these results in equation 1 you get

$$\boxed{\int \frac{1 - senx}{(1 + senx)senx} dx = ln\left| tan\frac{x}{2} \right| + \frac{4}{tan\left(\frac{x}{2}\right) + 1} + C}$$

237

$$\int \frac{dx}{\cos^2 x + 5\cos x + 6}$$

I watched the reader that the trigonometric function of the second degree is in the denominator of the integrand. Factoring the denominator creates two factors, which give us the idea of applying the technique of partial fraction integration.

$$\cos^2 x + 5\cos x + 6 = (\cos x + 3)(\cos x + 2)$$

$$\int \frac{dx}{\cos^2 x + 5\cos x + 6} = \int \frac{dx}{(\cos x + 3)(\cos x + 2)}$$

To solve the integral, the partial fraction integration method is applied.

$$\int \frac{dx}{(\cos x + 3)(\cos x + 2)} = \int \frac{A\,dx}{\cos x + 3} + \int \frac{B\,dx}{\cos x + 2}$$

$$1 = A(\cos x + 2) + B(\cos x + 3)$$
$$1 = A\cos x + 2A + B\cos x + 3B$$
$$1 = (A + B)\cos x + (2A + 3B)$$

$$\begin{vmatrix} A + B & = 0 \\ 2A + 3B & = 1 \end{vmatrix}$$ *Multiply the first equation by -2*

$$\begin{vmatrix} -2A - 2B & = 0 \\ 2A + 3B & = 1 \end{vmatrix}$$ $\rightarrow$ $\boldsymbol{B = 1}$ $\boldsymbol{A = -1}$

Consequently, you have

$$\int \frac{dx}{(\cos x + 3)(\cos x + 2)} = \int \frac{A\,dx}{\cos x + 3} + \int \frac{B\,dx}{\cos x + 2}$$

$$\int \frac{dx}{\cos^2 x + 5\cos x + 6} = -\int \frac{dx}{\cos x + 3} + \int \frac{dx}{\cos x + 2} \quad Ec.\,1$$

The integral is solved $\int \dfrac{dx}{\cos x+3}$

$$\int \frac{dx}{\cos x + 3} = \int \frac{\dfrac{2du}{1+u^2}}{\dfrac{1-u^2}{1+u^2}+3} = \int \frac{\dfrac{2du}{1+u^2}}{\dfrac{1-u^2+3+3u^2}{1+u^2}}$$

$$= 2\int \frac{du}{2u^2+4}$$

$$= 2\int \frac{du}{(\sqrt{2}\,u)^2+2^2}$$

$$= -\arctan\left(\frac{\sqrt{2}\,u}{2}\right)+C_1$$

$$\int \frac{dx}{\cos x+3} = -\arctan\left(\frac{\sqrt{2}\,u}{2}\right)+C_1$$

Solve the second integral of equation 1

$$\int \frac{dx}{\cos x + 2} = \int \frac{\dfrac{2du}{1+u^2}}{\dfrac{1-u^2}{1+u^2}+2} = \int \frac{\dfrac{2du}{1+u^2}}{\dfrac{1-u^2+2+2u^2}{1+u^2}}$$

$$= 2\int \frac{2\,du}{u^2+3}$$

$$= 2\int \frac{2\,du}{u^2+(\sqrt{3})^2}$$

$$= \frac{2}{\sqrt{3}}\arctan\left(\frac{u}{\sqrt{3}}\right)+C_2$$

$$\int \frac{dx}{\cos x+2} = \frac{2\sqrt{3}}{3}\arctan\left(\frac{u}{\sqrt{3}}\right)+C_2$$

The results of the integrals are substituted in equation 1

$$\int \frac{dx}{\cos^2 x + 5\cos x + 6} = -\arctan\left[\frac{1}{2}\tan\left(\frac{x}{2}\right)\sqrt{2}\right] + \frac{2}{3}\sqrt{3}\,\arctan\left[\frac{1}{3}\tan\left(\frac{x}{2}\right)\sqrt{3}\right] + C$$

238

$$\int \frac{6x - 1}{4x^2 + 4x + 10} dx$$

If the substitution or change of variable is not so recommendable, the viable alternative for the resolution of this integral is to simplify it, that is, to separate the integral into two integrals.

Equation 1

$$\int \frac{6x - 1}{4x^2 + 4x + 10} dx = 6 \int \frac{x}{4x^2 + 4x + 10} dx - \int \frac{dx}{4x^2 + 4x + 10}$$

The first integral is solved

$$6 \int \frac{x}{4x^2 + 4x + 10} dx \rightarrow u = 4x^2 + 4x + 10 \quad , \quad du = (8x + 4)dx$$

Multiply and divide the integral by 8 and add and subtract 4.

$$6 \int \frac{x}{4x^2 + 4x + 10} dx = \frac{6}{8} \int \frac{(8x + 4 - 4)dx}{4x^2 + 4x + 10}$$

$$= \frac{3}{4} \int \frac{(8x + 4)dx}{4x^2 + 4x + 10} - 3 \int \frac{dx}{4x^2 + 4x + 10}$$

This result is substituted in equation 1

$$\int \frac{6x - 1}{4x^2 + 4x + 10} dx = \frac{3}{4} \int \frac{(8x + 4)dx}{4x^2 + 4x + 10} - 3 \int \frac{dx}{4x^2 + 4x + 10} - \int \frac{dx}{4x^2 + 4x + 10}$$

$$\int \frac{6x - 1}{4x^2 + 4x + 10} dx = \frac{3}{4} \int \frac{(8x + 4)dx}{4x^2 + 4x + 10} - 4 \int \frac{dx}{4x^2 + 4x + 10} \quad \textbf{\textit{Equation 2}}$$

Solve the second integral of equation 2

$$4 \int \frac{dx}{4x^2 + 4x + 10}$$

It completes squares

$$4x^2 + 4x + 10 = 4\left[x^2 + x + \frac{10}{4}\right] = 4\left[x^2 + x + \frac{1}{4} + \frac{10}{4} - \frac{1}{4}\right]$$

$$= 4\left[\left(x + \frac{1}{2}\right)^2 + \frac{9}{4}\right] = 4\left[\left(x + \frac{1}{2}\right)^2 + \left(\frac{3}{2}\right)^2\right]$$

$$4\int \frac{dx}{4x^2 + 4x + 10} = \int \frac{dx}{\left(x + \frac{1}{2}\right)^2 + \left(\frac{3}{2}\right)^2}$$

This result is substituted in equation 2

$$\int \frac{6x - 1}{4x^2 + 4x + 10}\,dx = \frac{3}{4}\int \frac{(8x + 4)dx}{4x^2 + 4x + 10} - \int \frac{dx}{\left(x + \frac{1}{2}\right)^2 + \left(\frac{3}{2}\right)^2}$$

$$\int \frac{6x - 1}{4x^2 + 4x + 10}\,dx = \frac{3}{4}ln(4x^2 + 4x + 10) - \frac{2}{3}tan^{-1}\left(\frac{x + \frac{1}{2}}{\frac{3}{2}}\right) + C$$

$$\int \frac{6x - 1}{4x^2 + 4x + 10}\,dx = \frac{3}{4}ln(4x^2 + 4x + 10) - \frac{2}{3}tan^{-1}\left(\frac{2x + 1}{3}\right) + C$$

$$\int \frac{6x - 1}{4x^2 + 4x + 10}\,dx = \frac{3}{4}ln(2(2x^2 + 2x + 5)) - \frac{2}{3}tan^{-1}\left(\frac{2x + 1}{3}\right) + C$$

$$\int \frac{6x - 1}{4x^2 + 4x + 10}\,dx = \frac{3}{4}ln(2x^2 + 2x + 5) - \frac{2}{3}tan^{-1}\left(\frac{2x + 1}{3}\right) + C$$

$$\boxed{\int \frac{6x - 1}{4x^2 + 4x + 10}\,dx = \frac{3}{4}ln(2x^2 + 2x + 5) - \frac{2}{3}tan^{-1}\left(\frac{2}{3}x + \frac{1}{3}\right) + C}$$

239

$$\int \frac{(5 - 4x)dx}{\sqrt{12x - 4x^2 - 8}}$$

$$Sea \quad u = 12x - 4x^2 - 8 \quad , \quad du = (12 - 8x)dx$$

The reader can appreciate the difficulty that is presented so that the numerator of the integral is equal to du. To solve this situation, we proceed to divide the integral problem into two integrals, that is, rewrite it.

Equation 1

$$\int \frac{(5 - 4x)dx}{\sqrt{12x - 4x^2 - 8}} = 5\int \frac{dx}{\sqrt{12x - 4x^2 - 8}} - 4\int \frac{x\,dx}{\sqrt{12x - 4x^2 - 8}}$$

The first integral is solved

$$5\int \frac{dx}{\sqrt{12x - 4x^2 - 8}}$$

It completes squares

$$12x - 4x^2 - 8 = -4(x^2 - 3x + 2)$$

$$= -4\left[x^2 - 3x + \frac{9}{4} + 2 - \frac{9}{4}\right]$$

$$= -4\left[\left(x - \frac{3}{2}\right)^2 - \frac{1}{4}\right]$$

$$= -4\left[\left(x - \frac{3}{2}\right)^2 - \left(\frac{1}{2}\right)^2\right]$$

$$12x - 4x^2 - 8 = 4\left[\left(\frac{1}{2}\right)^2 - \left(x - \frac{3}{2}\right)^2\right]$$

$$5\int \frac{dx}{\sqrt{12x - 4x^2 - 8}} = 5\int \frac{dx}{\sqrt{4\left[\left(\frac{1}{2}\right)^2 - \left(x - \frac{3}{2}\right)^2\right]}}$$

$$5\int \frac{dx}{\sqrt{12x-4x^2-8}} = \frac{5}{2}sen^{-1}\left(\frac{x-\frac{3}{2}}{\frac{1}{2}}\right)+C_1$$

$$\boxed{5\int \frac{dx}{\sqrt{12x-4x^2-8}} = \frac{5}{2}sen^{-1}(2x-3)+C_1 \ \textbf{\textit{Equation 2}}}$$

Solve the second integral of equation 1

$$4\int \frac{x\,dx}{\sqrt{12x-4x^2-8}}$$

$$u = 12x-4x^2-8 \qquad du = (12-8x)dx$$

$$4\int \frac{x\,dx}{\sqrt{12x-4x^2-8}} = -\frac{4}{8}\int \frac{(-8x+12-12)\,dx}{\sqrt{12x-4x^2-8}}$$

$$4\int \frac{x\,dx}{\sqrt{12x-4x^2-8}} = -\frac{1}{2}\int \frac{(-8x+12)\,dx}{\sqrt{12x-4x^2-8}} + 6\int \frac{dx}{\sqrt{12x-4x^2-8}}$$

$$4\int \frac{x\,dx}{\sqrt{12x-4x^2-8}} = -\frac{1}{2}\int \frac{du}{\sqrt{u}} + 3\int \frac{dx}{\sqrt{\left[\left(\frac{1}{2}\right)^2-\left(x-\frac{3}{2}\right)^2\right]}}$$

$$\textbf{\textit{Equation 3}}$$

$$4\int \frac{x\,dx}{\sqrt{12x-4x^2-8}} = -\sqrt{12x-4x^2-8} + 3sen^{-1}(2x-3)+C_2$$

Comment: In the next step, common factor 4 is taken from the root and comes out as 2.

Substitute the results of equation 2 and equation 3 in equation 1

$$\int \frac{(5-4x)dx}{\sqrt{12x-4x^2-8}} = \frac{5}{2}sen^{-1}(2x-3)+2\sqrt{-x^2+3x-2}-3sen^{-1}(2x-3)+C$$

$$\boxed{\int \frac{(5-4x)dx}{\sqrt{12x-4x^2-8}} = -\frac{1}{2}sen^{-1}(2x-3)+2\sqrt{-x^2+3x-2}+C}$$

240

$$\int \sqrt{6x - x^2}\ dx$$

In the integrand of the integral problem we are presented with a root. In its subradical quantity, squares can be completed and then factorized. All this information leads us to the resolution of the integral proposal through an immediate integral.

$$6x - x^2 = -x^2 + 6x = -(x^2 - 6x)$$
$$6x - x^2 = -3[x^2 - 6x + 9 - 9]$$
$$6x - x^2 = -[(x - 3)^2 - 3^2]$$
$$6x - x^2 = 3^2 - (x - 3)^2$$

$$\int \sqrt{6x - x^2}\ dx = \int \sqrt{3^2 - (x - 3)^2}\ dx$$

$$\int \sqrt{6x - x^2}\ dx = \frac{x - 3}{2}\sqrt{9 - (x - 3)^2} + \frac{9}{2} sen^{-1}\left(\frac{x - 3}{3}\right) + C$$

The immediate integral is

$$\int \sqrt{a^2 - u^2}\ du = \frac{u}{2}\sqrt{a^2 - u^2} + \frac{a^2}{2} sen^{-1}\left(\frac{u}{a}\right) + C$$

$$\boxed{\int \sqrt{6x - x^2}\ dx = \frac{x - 3}{2}\sqrt{6x - x^2} + \frac{9}{2} sen^{-1}\left(\frac{1}{3}x - 1\right) + C}$$

241

$$\int e^{\sqrt{x}} dx$$

In the integral we are presented with the product of two functions. One exponential and the other algebraic. This suggests the application of the technique of integration by parts. Choose the easiest term to derive and dv the easiest term to integrate.

We will call $u = e^{\sqrt{x}}$ y $dv = dx$

$$u = e^{\sqrt{x}} \rightarrow du = e^{\sqrt{x}} \frac{1}{2\sqrt{x}} dx \rightarrow du = \frac{e^{\sqrt{x}} dx}{2\sqrt{x}}$$

$$dv = dx \rightarrow \int dv = \int dx \rightarrow v = x$$

Therefore, the integration by parts produces

$$\int e^{\sqrt{x}} dx = x \, e^{\sqrt{x}} - \int \frac{x \, e^{\sqrt{x}} dx}{2\sqrt{x}} \quad \textbf{\textit{Equation 1}}$$

The integral is solved

$$\int \frac{x \, e^{\sqrt{x}} dx}{2\sqrt{x}}$$

The integral is rewritten by changing the variable

$$w = \sqrt{x} \rightarrow dw = \frac{dx}{2\sqrt{x}}$$

$$\int \frac{x \, e^{\sqrt{x}} dx}{2\sqrt{x}} = \frac{1}{2} \int w^2 e^w \, dw$$

We will call $u = w^2$ and $dv = e^w dw$. According to the order of appearance in ALPES.

$$u = w^2 \rightarrow du = 2wdw$$

$$dv = e^w dw \rightarrow \int dv = \int e^w dw \rightarrow v = e^w$$

$$\frac{1}{2} \int w^2 e^2 \, dw = \frac{1}{2}\left[w^2 \, e^w - 2 \int w \, e^w dw \right]$$

$$\frac{1}{2} \int w^2 e^2 \, dw = \frac{1}{2} w^2 \, e^w - \int w \, e^w dw$$

This partial result is replaced in equation 1

$$\int e^{\sqrt{x}} dx = x \, e^{\sqrt{x}} - \int \frac{x \, e^{\sqrt{x}} dx}{2\sqrt{x}}$$

$$\int e^{\sqrt{x}} dx = x \, e^{\sqrt{x}} - \frac{1}{2} w^2 \, e^w + \int w \, e^w dw \quad \textbf{\textit{Equation 2}}$$

The integral is solved $\int w\ e^w dw$

We will call $u = w$ and $dv = e^w dw$. According to the order of appearance in ALPES.

$$u = w\ \rightarrow\ du = dw$$

$$dv = e^w dw\ \rightarrow \int dv = \int e^w dw\ \rightarrow\ v = e^w$$

$$\int w\ e^w dw = we^w - \int e^w dw\ \rightarrow\ \int w\ e^w dw = we^w - e^w + C$$

This partial result is replaced in equation 2

$$\int e^{\sqrt{x}} dx = x\ e^{\sqrt{x}} - \frac{1}{2} w^2\ e^w + we^w - e^w + C$$

$$\boxed{\int e^{\sqrt{x}} dx = x\ e^{\sqrt{x}} - \frac{1}{2} x\ e^{\sqrt{x}} + \sqrt{x}\ e^{\sqrt{x}} - e^{\sqrt{x}} + C}$$

242 $\qquad\qquad \boxed{\int (x^2 - x)e^{-x} dx}$

In the integral we are presented with the product of two functions. One algebraic and the other exponential. This suggests the application of the technique of integration by parts. Choose the easiest term to derive and dv the easiest term to integrate.

It is selecte $u = (x^2 - x)\ y\ dv = e^{-x}\ dx$

$$u = (x^2 - x)\ \rightarrow\ du = (2x - 1)dx$$

$$dv = e^{-x}\ dx \rightarrow \int dv =\ \rightarrow\ v = -e^{-x}$$

Therefore, the integration by parts produces

$$\int (x^2 - x)e^{-x} dx = -(x^2 - x)e^{-x} + \int (2x - 1)\ e^{-x} dx\ \textbf{\textit{Equation 1}}$$

The integral is solved $\int (2x - 1)\ e^{-x} dx$

$$u = (2x - 1)\ y\ dv = e^{-x}\ dx.$$

$$u = (2x - 1) \rightarrow du = 2dx$$

$$dv = e^{-x}\ dx\ \rightarrow \int dv = \int e^{-x}\ dx\ \rightarrow\ v = -e^{-x}$$

Therefore, the integration by parts produces

$$\int (2x-1)\,e^{-x}dx = -(2x-1)e^{-x} + 2\int e^{-x}dx \quad \textbf{\textit{Equation 2}}$$

The equation 2 is substituted in equation 1

$$\int (x^2-x)e^{-x}dx = -(x^2-x)e^{-x} - (2x-1)e^{-x} + 2\int e^{-x}dx$$

$$\int (x^2-x)e^{-x}dx = -(x^2-x)e^{-x} - (2x-1)e^{-x} - 2e^{-x} + C$$

$$\boxed{\int (x^2-x)e^{-x}dx = -e^{-x}(x^2+x+1) + C}$$

$$\int \frac{sen^2x\,dx}{e^x} \qquad\qquad \text{243}$$

The integral is rewritten by transforming the quotient into a product

$$\int \frac{sen^2x\,dx}{e^x} = \int sen^2x\,e^{-x}\,dx$$

Now we are presented with the product of two functions. The technique of integration by parts can be applied. You select the easiest factor to derive and the easiest to integrate. In our case: $= sen^2x \ y \ dv = e^{-x}$.

$$u = sen^2x \ \rightarrow \ du = 2sen\,x\cos x\,dx$$

$$dv = e^{-x}dx \rightarrow \int dv = \int e^{-x}\,dx$$

$$v = -e^{-x}$$

Therefore, the integration by parts produces

$$\int sen^2x\,e^{-x}\,dx = -e^{-x}\,sen^2x + 2\int e^{-x}\,sen\,x\cos x\,dx$$

As: $sen\,2x = 2sen\,x\cos x \rightarrow sen\,x\cos x = \frac{sen\,2x}{2}$

$$\int sen^2x\,e^{-x}\,dx = -e^{-x}\,sen^2x + 2\int \frac{sen\,2x}{2}e^{-x}\,dx$$

$$\int sen^2x\, e^{-x}\, dx = -e^{-x}\, sen^2x + \int sen\, 2x\, e^{-x}\, dx \quad \textbf{Equation 1}$$

Integration by parts is applied to the integral: $\int sen\, 2x\, e^{-x}\, dx$

$$w = sen\, 2x \rightarrow dw = 2\cos 2x\, dx$$

$$dv = e^{-x}dx \;\rightarrow\; \int dv = \int e^{-x}dx$$

$$v = -e^{-x}$$

Therefore, the integration by parts produces

$$\int sen\, 2x\, e^{-x}\, dx = -e^{-x}sen\, 2x + 2\int e^{-x}\cos 2x\, dx$$

This partial result is replaced in equation 1

$$\int sen^2x\, e^{-x}\, dx = -e^{-x}\, sen^2x - e^{-x}sen\, 2x + 2\int e^{-x}\cos 2x\, dx \quad \overset{\textbf{Equation 2}}{}$$

The integral is now resolved $\int e^{-x}\cos 2x\, dx$:

$$u = \cos 2x \rightarrow du = -2\, sen\, 2x\, dx$$

$$dv = e^{-x}dx \rightarrow \int dv = \int e^{-x}dx$$

$$v = -e^{-x}$$

Therefore, the integration by parts produces

$$\int e^{-x}\cos 2x\, dx = -e^{-x}\cos 2x - 2\int e^{-x}\, sen\, 2x\, dx$$

The integral is now resolved $\int e^{-x}\, sen\, 2x\, dx$:

$$u = sen\, 2x \;\rightarrow\; du = 2\cos 2x\, dx$$

$$dv = e^{-x}dx \rightarrow \int dv = \int e^{-x}dx \rightarrow v = -e^{-x}$$

$$\int e^{-x}\, sen\, 2x\, dx = -e^{-x}sen\, 2x + 2\int e^{-x}\cos 2x\, dx$$

$$\int e^{-x}\cos 2x\, dx = -e^{-x}\cos 2x - 2\left[-e^{-x}sen\, 2x + 2\int e^{-x}\cos 2x\, dx\right]$$

$$\int e^{-x} \cos 2x \, dx = -e^{-x} \cos 2x + 2e^{-x} sen\, 2x - 4 \int e^{-x} \cos 2x \, dx$$

$$\int e^{-x} \cos 2x \, dx + 4 \int e^{-x} \cos 2x \, dx = -e^{-x} \cos 2x + 2e^{-x} sen\, 2x$$

$$5 \int e^{-x} \cos 2x \, dx = -e^{-x} \cos 2x + 2e^{-x} sen\, 2x$$

$$\int e^{-x} \cos 2x \, dx = -\frac{1}{5} e^{-x} \cos 2x + \frac{2}{5} e^{-x} sen\, 2x$$

This partial result is replaced in equation 2

$$\int sen^2 x \, e^{-x} \, dx = -e^{-x} sen^2 x - e^{-x} sen\, 2x + 2 \left[-\frac{1}{5} e^{-x} \cos 2x + \frac{2}{5} e^{-x} sen\, 2x \right]$$

$$\int sen^2 x \, e^{-x} \, dx = -e^{-x} sen^2 x - e^{-x} sen\, 2x - \frac{2}{5} e^{-x} \cos 2x + \frac{4}{5} e^{-x} sen\, 2x$$

$$\int sen^2 x \, e^{-x} \, dx = -e^{-x} sen^2 x - \frac{1}{5} e^{-x} sen\, 2x - \frac{2}{5} e^{-x} \cos 2x$$

$$\boxed{\int sen^2 x \, e^{-x} \, dx = -e^{-x} sen^2 x - \frac{1}{5} e^{-x} (sen\, 2x + 2 \cos 2x) + c}$$

$$\boxed{\int sen^3 x \, \cos^2 x \, dx} \qquad\qquad \text{244}$$

It is a trigonometric integral that contains sine and cosine powers. The integration technique to solve is type of integrals is the following: As the power of the sine is odd and positive, the sine cube is decomposed and a sine is conserved. The other factors are converted to cosines.

$$\int sen^3 x \, \cos^2 x \, dx = \int sen^2 x \, \cos^2 x \, (sen\, x) \, dx$$

$$= \int (1 - \cos^2 x) \cos^2 x \, (sen\, x) \, dx)$$

$$= \int (\cos^2 x - \cos^4 x) \, (sen\, x) \, dx)$$

$$\int sen^3x \, cos^2x \, dx = \int cos^2x \, sen \, x \, dx - \int cos^4x \, sen \, x \, dx$$

Then the next substitution is made

$$u = cos \, x \rightarrow du = -sen \, x \, dx$$

$$\int sen^3x \, cos^2x \, dx = -\int u^2 du + \int u^4 du = -\frac{cos^3x}{3} + \frac{cos^5x}{5} + c$$

$$\boxed{\int sen^3x \, cos^2x \, dx = -\frac{cos^3x}{3} + \frac{cos^5x}{5} + c}$$

Notes

- The *senx* that is in parentheses in the first step is the factor that is conserved.
- To convert the other factors to cosines, the Pythagorean trigonometric identity is used $sen^2x + cos^2x = 1$.

245 ━━━━━━━━━━━━━━━━━━━━━━━━━━ $\boxed{\displaystyle\int tan^3 \frac{\pi x}{2} \, sec^2 \frac{\pi x}{2} dx}$ ━━━━━

The theory says that if the power of the secant is even and positive, keep a square secant factor and convert the remaining factors into tangents. As you can see, the integral presents a square secant factor, and there are no remaining factors to turn them into tangents. What is done then? What is appropriate is to convert the square secant into a square tangent.

$$\int tan^3 \left(\frac{\pi x}{2}\right) sec^2 \left(\frac{\pi x}{2}\right) dx = \int tan^3 \left(\frac{\pi x}{2}\right) \left[1 + tan^2 \left(\frac{\pi x}{2}\right)\right] dx$$

$$\int tan^3 \left(\frac{\pi x}{2}\right) sec^2 \left(\frac{\pi x}{2}\right) dx = \int tan^3 \frac{\pi x}{2} dx + \int tan^5 \frac{\pi x}{2} dx$$

Then we proceed to solve each integral, and adding the results we get the result of the integral problem.

$$\int tan^3 \left(\frac{\pi x}{2}\right) dx = tan^2 \left(\frac{\pi x}{2}\right) tan \left(\frac{\pi x}{2}\right) dx$$

$$= \int \left[sec^2 \left(\frac{\pi x}{2} \right) - 1 \right] tan \left(\frac{\pi x}{2} \right) dx$$

$$= \int tan \left(\frac{\pi x}{2} \right) sec^2 \left(\frac{\pi x}{2} \right) dx - \int tan \left(\frac{\pi x}{2} \right) dx$$

The following substitution is made $u = tan\frac{\pi x}{2} \rightarrow du = \frac{\pi}{2} sec^2 \frac{\pi x}{2} dx$

$$\int tan^3 \frac{\pi x}{2} dx = \frac{\pi}{2} \int u du - \int tan\frac{\pi x}{2} dx$$

$$\int tan^3 \frac{\pi x}{2} dx = \frac{2}{\pi} \frac{tan^2 \frac{\pi x}{2}}{2} - \frac{2}{\pi} ln \left| sec \frac{\pi x}{2} \right| + C$$

$$\boxed{\int tan^3 \frac{\pi x}{2} dx = \frac{1}{\pi} tan^2 \frac{\pi x}{2} - \frac{2}{\pi} ln \left| sec \frac{\pi x}{2} \right| + C}$$

$$\int tan^5 \frac{\pi x}{2} dx = \int tan^2 \frac{\pi x}{2} tan^3 \frac{\pi x}{2} dx$$

$$= \int \left(sec^2 \frac{\pi x}{2} - 1 \right) tan^3 \frac{\pi x}{2} dx$$

$$= \int tan^3 \frac{\pi x}{2} sec^2 \frac{\pi x}{2} dx - \int tan^3 \frac{\pi x}{2} dx$$

$$= \int tan^3 \frac{\pi x}{2} sec^2 \frac{\pi x}{2} dx - \int \left(sec^2 \frac{\pi x}{2} - 1 \right) tan \frac{\pi x}{2} dx$$

$$= \int tan^3 \frac{\pi x}{2} sec^2 \frac{\pi x}{2} dx - \int tan \frac{\pi x}{2} sec^2 \frac{\pi x}{2} dx + \int tan \frac{\pi x}{2} dx$$

The following substitution is made $u = tan\frac{\pi x}{2} \rightarrow du = \frac{\pi}{2} sec^2 \frac{\pi x}{2} dx$

$$\int tan^5 \frac{\pi x}{2} dx = \frac{2}{\pi} \int u^3 du - \frac{2}{\pi} \int u du + \int tan \frac{\pi x}{2} dx$$

$$\int tan^5 \frac{\pi x}{2} dx = \frac{2}{\pi} \frac{tan^4 \frac{\pi x}{2}}{4} - \frac{2}{\pi} \frac{tan^2 \frac{\pi x}{2}}{2} + \frac{2}{\pi} ln \left| sec \frac{\pi x}{2} \right| + c$$

Consequently, you have

$$\int \tan^3 \frac{\pi x}{2} \sec^2 \frac{\pi x}{2}\, dx = \int \tan^3 \frac{\pi x}{2}\, dx + \int \tan^5 \frac{\pi x}{2}\, dx$$

$$\int \tan^3 \frac{\pi x}{2} \sec^2 \frac{\pi x}{2}\, dx = \frac{1}{\pi}\tan^2 \frac{\pi x}{2} - \frac{2}{\pi}\ln\left|\sec \frac{\pi x}{2}\right| + c + \;\longrightarrow$$

$$\longrightarrow \; +\frac{1}{2\pi}\tan^4 \frac{\pi x}{2} - \frac{1}{\pi}\tan^2 \frac{\pi x}{2} + \frac{2}{\pi}\ln\left|\sec \frac{\pi x}{2}\right| + c$$

Eliminating similar terms you get

$$\boxed{\int \tan^3 \frac{\pi x}{2} \sec^2 \frac{\pi x}{2}\, dx = \frac{1}{2\pi}\tan^4 \frac{\pi x}{2} + c}$$

246

$$\int \frac{dx}{\tan x + sen\, x}$$

It is an integral that contains two trigonometric functions: a tangent and a sine. If the tangent is expressed in terms of sine and cosine, a sine and cosine integral is obtained.

$$\int \frac{dx}{\tan x + sen\, x} = \int \frac{dx}{\frac{sen\, x}{\cos x} + sen\, x}$$

For the resolution of integrals containing sine and cosine functions, the following substitutions will be made.

$$sen\, x = \frac{2u}{1+u^2}\;,\quad \cos x = \frac{1-u^2}{1+u^2}\;,\quad dx = \frac{2}{1+u^2}du\;,\quad u = \tan \frac{x}{2}$$

The quotient of the sen x y el cos x.

$$\frac{sen\, x}{\cos x} = \frac{\frac{2u}{1+u^2}}{\frac{1-u^2}{1+u^2}} = \frac{2u}{1-u^2}$$

$$\int \frac{dx}{\frac{sen\,x}{cos\,x} + sen\,x} = \int \frac{\frac{2}{1+u^2}du}{\frac{2u}{1-u^2} + \frac{2u}{1+u^2}} = \int \frac{\frac{2}{1+u^2}du}{\frac{2u(1+u^2) + 2u(1-u^2)}{(1-u^2)(1+u^2)}}$$

$$\int \frac{dx}{\frac{sen\,x}{cos\,x} + sen\,x} = \int \frac{\frac{2}{1+u^2}du}{\frac{2u + 2u^3 + 2u - 2u^3}{(1-u^2)(1+u^2)}} = \int \frac{\frac{2}{(1+u^2)}du}{\frac{4u}{(1-u^2)(1+u^2)}}$$

$$\int \frac{dx}{\frac{sen\,x}{cos\,x} + sen\,x} = \int \frac{2}{\frac{4u}{(1-u^2)}}du = \int \frac{2 - 2u^2}{4u}du$$

$$\int \frac{dx}{\frac{sen\,x}{cos\,x} + sen\,x} = \int \frac{2(1-u^2)}{4u}du = \frac{1}{2}\int \frac{1-u^2}{u}du$$

$$\int \frac{dx}{\frac{sen\,x}{cos\,x} + sen\,x} = -\frac{1}{2}\int \frac{u^2-1}{u}du \;\rightarrow$$
It multiplied by minus (-) the integral to change position the terms of the numerator.

$$\int \frac{dx}{\frac{sen\,x}{cos\,x} + sen\,x} = -\frac{1}{2}\int \left(u - \frac{1}{u}\right)du \;\longrightarrow$$

$$\int \frac{dx}{\frac{sen\,x}{cos\,x} + sen\,x} = -\frac{1}{2}\int u\,du + \frac{1}{2}\int \frac{du}{u}$$

$$\int \frac{dx}{\frac{sen\,x}{cos\,x} + sen\,x} = -\frac{1}{4}u^2 + \frac{1}{2}ln|u| + C$$

$$\boxed{\int \frac{dx}{\frac{sen\,x}{cos\,x} + sen\,x} = \frac{1}{2}ln\left|\tan\frac{x}{2}\right| - \frac{1}{4}tan^2\frac{x}{2} + C}$$

247

$$\int \frac{\sqrt{1-x}}{\sqrt{x}}\, dx$$

The integral contains a radical of the form $\sqrt{1-x}$, to transform it to the form $\sqrt{a^2 - u^2}$ he applies the following mathematical artifice: $\left(\sqrt{x}\right)^2$ and the 1 en 1^2. That is to say, $\sqrt{(1)^2 - \left(\sqrt{x}\right)^2}$. The previous thing leads us to the application of the integration technique by trigonometric substitution. Therefore, we have:

$$\int \frac{\sqrt{1-x}}{\sqrt{x}}\, dx = \int \frac{\sqrt{1^2 - (\sqrt{x})^2}}{\sqrt{x}}\, dx$$

$$a = 1, \sqrt{x} = sen\theta \;\rightarrow\; x = sen^2\theta \;\rightarrow\; dx = 2sen\theta\, cos\theta\, d\theta, \sqrt{1^2 - (\sqrt{x})^2} = cos\theta$$

Now we proceed to make the substitution and solve the integral

$$\int \frac{\sqrt{1^2 - (\sqrt{x})^2}}{\sqrt{x}}\, dx = \int \frac{cos\theta}{sen\theta}\, 2sen\theta\, cos\theta\, d\theta = 2\int cos^2\theta\, d\theta$$

$$= \int (1 + cos2\theta)\, d\theta = \int d\theta + \int cos2\theta\, d\theta$$

$$= \theta + \frac{1}{2}\, sen2\theta + C = \theta + \frac{1}{2}\, 2sen\theta cos\theta + C$$

$$\int \frac{\sqrt{1^2 - (\sqrt{x})^2}}{\sqrt{x}}\, dx = \theta + sen\theta cos\theta + C$$

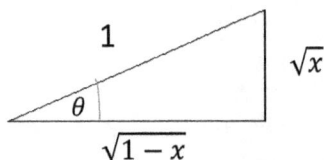

$$sen\theta = \frac{\sqrt{x}}{1}, \; \theta = arc\, sen\sqrt{x}$$

$$\boxed{\int \frac{\sqrt{1-x}}{\sqrt{x}}\, dx = arc\, sen\sqrt{x} + \sqrt{x}\,\sqrt{1-x} + C}$$

$$\int \frac{dx}{x - \sqrt[3]{x}}$$

248

The integral contains an irrational expression $\sqrt[3]{x}$, which suggests the application of the technique of integration of irrational functions. Therefore, the following variable change is made. $x = z^n$

Where n is the index of the root. $\quad x = z^3 \longrightarrow dx = 3z^2 dz, \ \sqrt[3]{z} = z$

Therefore, you have

$$\int \frac{dx}{x - \sqrt[3]{x}} = \int \frac{3z^2 dz}{z^3 - z} = 3 \int \frac{z^2 dz}{z(z^2 - 1)}$$

$$\int \frac{dx}{x - \sqrt[3]{x}} = 3 \int \frac{z \, dz}{z^2 - 1}$$

Where: $t = z^2 - 1 \quad dt = 2zdz$

$$\int \frac{dx}{x - \sqrt[3]{x}} = \frac{3}{2} \int \frac{dt}{t}$$

$$\int \frac{dx}{x - \sqrt[3]{x}} = \frac{3}{2} \ln|t| + C$$

$$\int \frac{dx}{x - \sqrt[3]{x}} = \frac{3}{2} \ln|z^2 - 1| + C$$

$$\boxed{\int \frac{dx}{x - \sqrt[3]{x}} = \frac{3}{2} \ln \left| x^{\frac{2}{3}} - 1 \right| + C}$$

$$\int \frac{dx}{x(\sqrt{x-1}-1)}$$

Case analogous to the previous exercise

$$x - 1 = z^2 \quad \rightarrow \quad x = z^2 + 1 \quad \rightarrow \quad dx = 2z\,dz \quad \rightarrow \quad z = \sqrt{x-1}$$

$$\int \frac{dx}{x(\sqrt{x-1}-1)} = 2\int \frac{z\,dz}{(z^2+1)(z-1)}$$

The resulting integral is solved by applying the partial fraction technique.

$$\frac{z\,dz}{(z^2+1)(z-1)} = \frac{Az+B}{z^2+1} + \frac{C}{z-1}$$

$$z = (Az+B)(z-1) + C(z^2+1)$$

$$z = Az^2 - Az + Bz - B + Cz^2 + C$$

$$z = (A+C)z^2 + (-A+B)z + (-B+C)$$

From the above, the following system of equations originates.

$$
\begin{array}{rrrl}
A & & + C & = 0 \\
-A & + B & & = 1 \\
& -B & + C & = 0
\end{array}
$$

The system of equations is solved and the following values are obtained.

$$A = -\frac{1}{2}, \qquad B = \frac{1}{2} \qquad and \qquad C = \frac{1}{2}$$

Consequently, you have

$$\int \frac{z\,dz}{(z^2+1)(z-1)} = \int \frac{\left(-\frac{1}{2}z + \frac{1}{2}\right)dz}{z^2+1} + \int \frac{\frac{1}{2}dz}{z-1}$$

$$\int \frac{z\,dz}{(z^2+1)(z-1)} = -\frac{1}{2}\int \frac{z\,dz}{z^2+1} + \frac{1}{2}\int \frac{dz}{z^2+1} + \frac{1}{2}\int \frac{dz}{z-1}$$

$$\int \frac{z\,dz}{(z^2+1)(z-1)} = -\frac{1}{4}\int \frac{du}{u} + \frac{1}{2}\int \frac{dz}{z^2+1} + \frac{1}{2}\int \frac{dt}{t}$$

$$\int \frac{z\,dz}{(z^2+1)(z-1)} = -\frac{1}{4}\ln|u| + \frac{1}{2}tan^{-1}z + \frac{1}{2}\ln|t| + C$$

$$\int \frac{z\,dz}{(z^2+1)(z-1)} = -\frac{1}{4}\ln|z^2+1| + \frac{1}{2}tan^{-1}z + \frac{1}{2}\ln|z-1| + C$$

$$\int \frac{z\,dz}{(z^2+1)(z-1)} = -\frac{1}{4}\ln|x-1+1| + \frac{1}{2}tan^{-1}\left(\sqrt{x-1}\right) + \frac{1}{2}\ln|\sqrt{x-1}-1| + C$$

$$\int \frac{z\,dz}{(z^2+1)(z-1)} = -\frac{1}{4}\ln|x| + \frac{1}{2}tan^{-1}\left(\sqrt{x-1}\right) + \frac{1}{2}\ln|\sqrt{x-1}-1| + C$$

$$\int \frac{z\,dz}{(z^2+1)(z-1)} = \frac{1}{2}\left[-\frac{1}{2}\ln|x| + tan^{-1}\left(\sqrt{x-1}\right) + \ln|\sqrt{x-1}-1|\right] + C$$

$$\int \frac{z\,dz}{(z^2+1)(z-1)} = 2*\frac{1}{2}\left[\ln|\sqrt{x-1}-1| - \frac{1}{2}\ln|x| + tan^{-1}\left(\sqrt{x-1}\right)\right] + C$$

$$\int \frac{z\,dz}{(z^2+1)(z-1)} = \left[\ln|\sqrt{x-1}-1| - \frac{1}{2}\ln|x| + tan^{-1}\left(\sqrt{x-1}\right)\right] + C$$

$$\boxed{\int \frac{dx}{x(\sqrt{x-1}-1)} = \ln\left|\frac{\sqrt{x-1}-1}{\sqrt{x}}\right| + tan^{-1}\left(\sqrt{x-1}\right) + C}$$

250

$$\int \frac{x^2\,dx}{\sqrt{21 + 4x - x^2}}$$

The reader must always bear in mind the importance of extracting the necessary information provided by the integral for its resolution. In our case, the integral presents a radical in the denominator, its subradical quantity is a function of the second degree. By completing squares you can transform the radical to the form $\sqrt{a^2 - u^2}$. And apply the trigonometric substitution technique.

It completes squares to transform the radical to the form $\sqrt{a^2 - u^2}$.

$$21 + 4x - x^2 = -(x^2 - 4x - 21) = -(x^2 - 4x + 4) - 21 - 4$$

$$21 + 4x - x^2 = -[(x - 2)^2 - 25] = 25 - (x - 2)^2 = 5^2 - (x - 2)^2$$

$$\int \frac{x^2\,dx}{\sqrt{21 + 4x - x^2}} = \int \frac{x^2\,dx}{\sqrt{5^2 - (x - 2)^2}}$$

$a = 5,\ x - 2 = 5sen\theta,\ x = 5sen\theta + 2,\ dx = 5cos\theta d\theta,\ \sqrt{5^2 - (x - 2)^2} = 5cos\theta$

$$\int \frac{x^2\,dx}{\sqrt{5^2 - (x - 2)^2}} = \int \frac{(5sen\theta + 2)^2 5cos\theta d\theta}{5cos\theta}$$

$$= \int \frac{(5sen\theta + 2)^2 5cos\theta d\theta}{5cos\theta}$$

$$= \int (5sen\theta + 2)^2$$

$$= \int (25sen^2\theta + 20sen\theta + 4)\,d\theta$$

$$= 25\int sen^2\theta d\theta + 20\int sen\theta d\theta + 4\int d\theta$$

$$= \frac{25}{2}\int (1 - cos2\theta)d\theta + 20\int sen\theta d\theta + 4\int d\theta$$

$$= \frac{25}{2} \int d\theta - \frac{25}{2} \int cos2\theta d\theta + 20 \int sen\theta d\theta + 4 \int d\theta$$

$$= \frac{25}{2} \theta - \frac{25}{4} sen2\theta - 20cos\theta + 4\theta + C$$

$$= \frac{33}{2} \theta - \frac{25}{2} sen\theta cos\theta - 20cos + C$$

Remember: $sen^2\theta = \dfrac{(1 - cos2\theta)}{2}$

Remember: $sen2\theta = 2sen\theta cos\theta$

$$\int \frac{x^2 dx}{\sqrt{5^2 - (x - 2)^2}} = \frac{33}{2} \theta - \frac{25}{2} sen\theta cos\theta - 20cos + C$$

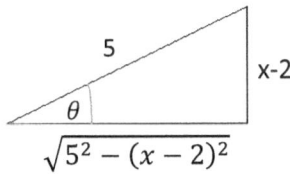

$$x - 2 = 5sen\theta \rightarrow sen\theta \frac{x - 2}{5}$$

$$\theta = arcsen \frac{x - 2}{5}$$

$$\int \frac{x^2 dx}{\sqrt{21 + 4x - x^2}} = \frac{33}{2} arcsen\frac{x - 2}{5} - \frac{25}{2} \frac{(x - 2)}{5} \frac{\sqrt{25 - (x - 2)^2}}{5} - 20\frac{\sqrt{25 - (x - 2)^2}}{5} + C$$

$$\int \frac{x^2 dx}{\sqrt{21 + 4x - x^2}} = \frac{33}{2} arcsen\frac{x - 2}{5} - \frac{1}{2} (x - 2)\sqrt{21 + 4x - x^2} - 4\sqrt{21 + 4x - x^2} + C$$

$$\int \frac{x^2 dx}{\sqrt{21 + 4x - x^2}} = \frac{33}{2} arcsen\frac{x - 2}{5} - \sqrt{21 + 4x - x^2} \left[\frac{1}{2}(x - 2) + 4\right] + C$$

$$\int \frac{x^2 dx}{\sqrt{21 + 4x - x^2}} = \frac{33}{2} arcsen\frac{x - 2}{5} - \sqrt{21 + 4x - x^2} \left(\frac{x + 6}{2}\right) + C$$

$$\boxed{\int \frac{x^2 dx}{\sqrt{21 + 4x - x^2}} = \frac{33}{2} arcsen\frac{x - 2}{5} - \sqrt{21 + 4x - x^2} \left(\frac{x + 6}{2}\right) + C}$$

EDITORIAL
INFINITO
[∞]